Lecture Notes in Artificial Inte

Edited by R. Goebel, J. Siekmann, and W. Wahlster

Subseries of Lecture Notes in Computer Science

Nuno David   Jaime Simão Sichman (Eds.)

Multi-Agent-Based Simulation IX

International Workshop, MABS 2008
Estoril, Portugal, May 12-13, 2008
Revised Selected Papers

 Springer

Series Editors

Randy Goebel, University of Alberta, Edmonton, Canada
Jörg Siekmann, University of Saarland, Saarbrücken, Germany
Wolfgang Wahlster, DFKI and University of Saarland, Saarbrücken, Germany

Volume Editors

Nuno David
ISCTE - Lisbon University Institute
Av. das Forças Armadas
1649-026 Lisboa, Portugal
E-mail: nuno.david@iscte.pt

Jaime Simão Sichman
University of São Paulo
Computer Engineering Department
Av. Prof. Luciano Gualberto, tv.3, 158
05508-970 São Paulo SP Brazil
E-mail: jaime.sichman@poli.usp.br

Library of Congress Control Number: Applied for

CR Subject Classification (1998): I.2, I.2.11, J.4, K.4

LNCS Sublibrary: SL 7 – Artificial Intelligence

ISSN   0302-9743
ISBN-10   3-642-01990-0 Springer Berlin Heidelberg New York
ISBN-13   978-3-642-01990-6 Springer Berlin Heidelberg New York

springer.com

© Springer-Verlag Berlin Heidelberg 2009
Printed in Germany

Typesetting: Camera-ready by author, data conversion by Scientific Publishing Services, Chennai, India
Printed on acid-free paper   SPIN: 12675531   06/3180   5 4 3 2 1 0

Preface

The meeting of researchers from multi-agent systems engineering and the social/economic/organizational sciences plays a vital role in the cross-fertilization of ideas, and is undoubtedly an important source of inspiration for the body of knowledge that is being produced in the multi-agent field. The MABS series continues to pursue its goal of bringing together researchers interested in multi-agent systems with those focused on modelling complex social systems, in such areas as economics, management, and organizational and social sciences in general.

This volume is the ninth of its series. It is based on papers accepted for the 9th International Workshop on Multi-agent-Based Simulation (MABS 2008), co-located with the 7th International Joint Conference on Autonomous Agents and Multiagent Systems (AAMAS 2008), which was held in Estoril, Portugal, May 12–16, 2008. All the papers presented at the workshop have been extended, revised and reviewed again in order to be part of this volume. The success of this field is reflected in the outstanding number of submissions that we received at that time. Forty-four submissions from 14 countries were received, from which we selected 16 for presentation (near 35% acceptance). We are very grateful to the participants who provided a lively atmosphere of debate during the presentation of the papers and during the general discussion on the challenges that the MABS field faces.

We are also very grateful to all the members of the Program Committee and the additional reviewers for their hard work. Thanks are also due to Juan A. Rodriguez-Aguilar (AAMAS 2008 Workshop Chair), to Simon Parsons and Joerg P. Mueller (AAMAS 2008 General Chairs), to Lin Padgham and David Parkes (AAMAS 2008 Program Chairs) and to Ana Paiva and Luis Antunes (AAMAS 2005 Local Organization Chairs).

March 2009
Nuno David
Jaime Sichman

Organization

General and Program Chairs

Nuno David   Lisbon University Institute-ISCTE, Portugal
Jaime Simão Sichman   University of São Paulo, Brazil

Program Committee

Adolfo López Paredes	INSISOC, Valladolid, Spain
Akira Namatame	National Defense Academy, Japan
Alexis Drogoul	IRD, MSI Research Team, Vietnam
Ana Bazzan	Federal University of Rio Grande do Sul, Brazil
Carles Sierra	IIIA, Spain
Cesáreo Hernández Iglesias	INSISOC, Valladolid, Spain
Claudio Cioffi-Revilla	George Mason University, USA
Cristiano Castelfranchi	ISTC/CNR, Italy
David Hales	University of Bologna, Italy
David Sallach	Argonne National Lab and University of Chicago, USA
Diana Adamatti	University of São Paulo, Brazil
Elizabeth Sklar	City University of New York, USA
Emma Norling	Manchester Metropolitan University, UK
Ernesto Costa	University of Coimbra, Portugal
Fréedéric Amblard	University of Toulouse, France
H. Van Parunak	NewVectors LLC, USA
Harko Verhagen	Stockholm University, Sweden
Helder Coelho	Lisbon University, Portugal
Jaime Sichman	University of São Paulo, Brazil (PC Co-chair)
Jan Treur	Vrije Universiteit in Amsterdam, The Netherlands
João Balsa	Universidade de Lisboa, Portugal
Jorge Louçã	ISCTE, Portugal
Juan Pavon Mestras	Universidad Complutense Madrid, Spain
Juliette Rouchier	Greqam/CNRS, France
Keith Sawyer	Washington University in St. Louis, USA
Keiki Takadama	University of Electro-Communications, Japan
Klaus Troitzsch	University of Koblenz, Germany
Liz Sonenberg	University of Melbourne, Australia
Luis Antunes	University of Lisbon, Portugal

Marco Janssen	Indiana University, USA
Maria Marietto	Universidade Federal do ABC, Brazil
Mario Paolucci	IP/CNR Rome, Italy
Nick Gotts	Macaulay Institute, UK
Nigel Gilbert	University of Surrey, UK
Nuno David	Lisbon University Institute, ISCTE, Portugal (PC Co-chair)
Oswaldo Teran	University of Los Andes, Venezuela
Paul Davidsson	Blekinge Institute of Technology, Sweden
Paulo Novais	Universidade do Minho, Portugal
Rainer Hegselmann	University of Bayreuth, Germany
Robert Axtell	George Mason University, USA
Rosaria Conte	ISTC/CNR Rome, Italy
Satoshi Kurihara	Osaka University, Japan
Scott Moss	Manchester Metropolitan University, UK
Sung-Bae Cho	Yonsei University, Korea
Takao Terano	University of Tsukuba, Japan
Wander Jager	University of Groningen, The Netherlands

Additional Referees

Alexei Sharpanskykh, The Netherlands
Anarosa Brandão, Brazil
Charlotte Gerritsen, The Netherlands
Fiemke Both, The Netherlands
José Eurico Filho, Brazil
Luís Mota, Portugal
Marc Esteva, Spain

Table of Contents

Simulation of Economic Behaviour

Modelling and Simulation of Social Behaviour

Applications

Techniques, Infrastructure and Technologies

Methods and Methodologies

Modeling Power Distance in Trade

Gert Jan Hofstede[1], Catholijn M. Jonker[2], and Tim Verwaart[3]

[1] Wageningen University, Postbus 9109, 6700 HB Wageningen, The Netherlands
gertjan.hofstede@wur.nl
[2] Delft University of Technology, Mekelweg 4, 2628 CD Delft, The Netherlands
c.m.jonker@tudelft.nl
[3] LEI Wageningen UR, Postbus 29703, 2502 LS den Haag, The Netherlands
tim.verwaart@wur.nl

Abstract. Agent-based computational economics studies the nature of economic processes by means of artificial agents that simulate human behavior. Human behavior is known to be scripted by cultural background. The processes of trade partner selection and negotiation work out differently in different communities. Different communities have different norms regarding trust and opportunism. These differences are relevant for processes studied in economics, especially for international trade. This paper takes Hofstede's model of national culture as a point of departure. It models the effects on trade processes of one of the five dimensions: power distance. It formulates rules for the behavior of artificial trading agents and presents a preliminary verification of the rules in a multi-agent simulation.

Keywords: culture, negotiation, trust, deceit, simulation.

1 Introduction

Any experienced international traveler knows that economic transactions do not come to pass in the same way across cultures. Haggling, checking on quality, and style of negotiation vary considerably across the world.

In the quest to understand the mechanisms that underlie these differences this article adopts the approach of designing agent-based simulation models. It builds on [1], that describes the modeling of behavioral differences of participants in a human gaming simulation. The game gives players the choice to either trust their trade partners to live up to their promises, or to spend money, time, and relational assets to check (trace) them. In the game, differences are observed between players from different cultural backgrounds [2]. Generally negotiation - which is an essential process in trade - is recognized to develop differently in different cultural settings, see e.g. [3]. For electronically mediated negotiations, [4] reports considerable differences across countries with respect to expectations and process.

Negotiation relates to the pre-contract phase of economic transactions. Trust and opportunism predominantly relate to the post-contract phase: the delivery. [5] gives evidence that both trust and opportunism can be profitable in this phase. It suggests

N. David and J.S. Sichmann (Eds.): MABS 2008, LNAI 5269, pp. 1–16, 2009.

that in different societies self-sustaining systems of either trust or opportunism might prevail. [6] supports these findings: the extent to which people expect deceit and are likely to lie in business negotiations differs considerably across cultures.

The discipline of agent-based economics [7] recognizes that using artificial agents to simulate human behavior contributes to the understanding of economic processes. Models of cultural influences on behavior in searching, bargaining, monitoring, and enforcing contracts are essential for developing realistic agents that can help us understand the differentiation of economic systems and institutions across the world. The design of culturally scripted agents serves several purposes. First it is useful for research into the effects of culture in trade, as described in the previous paragraph. Secondly, it can be used in education and training to make traders aware of cultural differences. Furthermore, the models can be used for developing negotiation support systems.

The approach taken by the authors is to make use of the widely used 5-dimension framework of Hofstede [8]. The present paper's research goal is to investigate the role of the cultural dimension of power distance as a determinant of trade processes and outcomes. We adopt the perspective of the trader that uses the endemic logic of a particular orientation on the power distance scale.

2 Power Distance and Trade

Can traders predict the behavior of potential partners depending on which part of the world these partners come from? Granting that each individual is unique, they can. For this, traders need knowledge about the socialization that the potential partners underwent in childhood, in other words about their culture. In many cases, nationality is a good predictor of the participants' basic values. For instance, business in China tends to be done over a meal, and observing social hierarchy during meals is important. In the Netherlands, business is done during working hours and little concern is given to the formal status of traders. This statement is inadequate for some Chinese and some Dutch traders but it is certainly more true than its opposite would be. The work of Hofstede [8, 9] characterizes these values in the form of five basic dimensions of social life that pertain to identity, power distance, gender roles, fear of the unknown, and long- vs short-term orientation.

The dimension of power distance is central in the present paper. Hofstede [8] defines power distance as the extent to which the less powerful accept and expect that power is distributed unequally. The dimension runs from egalitarian (*small power distance, e.g., in* Anglo, Germanic and Nordic cultures) to hierarchical (*large power distance,* in most other cultures; see table 1).

Table 1. Some distinctions between norms in hierarchical and egalitarian societies

Large power distance (hierarchical)	Small power distance (egalitarian)
Might is right	No privileges and status symbols
Formal speech; acknowledgement	Talk freely in any context
Dictate, obey	Negotiate
Show favor to mighty business partners	Treat all business partners equally

There are some pairs of countries in the Hofstede database that differ on power distance more than they do on other dimensions. They are Russia - Israel, Costa Rica - Guatemala, and France - Austria. Still it would not do to take subjects from these pairs of countries, have them negotiate, and attribute the results to difference in power distance. Besides cultural differences on other dimensions, differences in perceived identity, historical antecedents, personality factors and a host of other context factors have to be taken into account. By using software agents, these other contextual factors can be excluded and power distance can be isolated. However, simulation results will have to be interpreted as an abstraction that cannot be extrapolated to the real world without much caution. Isolating one dimension for the sake of experiment is a decidedly artificial method. In real life, the dimensions always operate as one whole, a cultural Gestalt, together with contextual factors. One of the contextual factors is personality: in any trade situation it matters what personalities the partners bring to the table. As it turns out personality and culture are not independent. In a meta-analysis of their mutual cross-country data Hofstede and McCrae [10] found that power distance correlates negatively with extraversion and openness to ideas and positively with conscientiousness.

In spite of the limitations of isolating a single dimension, we argue that the experiment is worthwhile carrying out. Empiric evidence for the relevance of the power distance dimension for negotiation processes is given in [3]. Furthermore, modeling the isolated dimensions can serve as a preparation for the more complicated integral modeling of culture's consequences for trade.

The core of trade is the execution of transactions: exchanging commodities or rights for money. Transactions are based on a contract that may specify additional conditions that enforce the delivery according to the contract. The contract is to be negotiated among the trade partners. Contracting is not the only relevant activity of trading agents, however. Fig. 1 presents an overview of relevant processes.

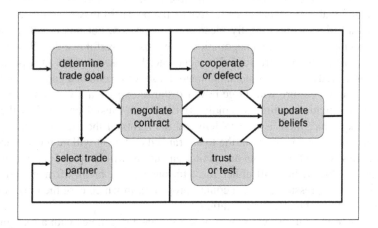

Fig. 1. Processes and internal information flow of a trading agent

Before entering negotiations, agents have to select each other as partners to negotiate with, based on their trade goals (sell or buy?; which commodity?; which quality?) and knowledge about potential partners. This information also plays an important role during the process of negotiation. Once a contract has been agreed upon, the traders can either cooperate (deliver truthfully) or try and use an opportunity to defect. Upon delivery the receiver can either trust or put the delivery to the test, the latter usually at some cost where trust is for free. The delivery and trust decisions are based on personal preferences and cultural background, as well as on beliefs about the trade partner and the trade environment. Experience from the negotiation and delivery processes may change a trader's beliefs about the trade environment or about individual trade partners. The beliefs will, in addition to a trader's preferences, guide decision making in future trading. In this paper we limit ourselves to beliefs about trade partners. For this purpose three traits can be defined that trading agents maintain a belief about: fairness, trustworthiness, and benevolence. These beliefs are maintained for each trade partner. The belief about another agent's *fairness* represents an agent's expectation that a fair contract can be negotiated with the other agent. The belief about another agent's *trustworthiness* represents an agent's expectation that the other agent will deliver according to contract. The belief about another agent's *benevolence* represents the expectation that the other agent will accept deliveries without putting them to the test, in other words that the other agent will trust.

The power distance dimension has its effect on behavior in trade. The relevant issues are a trader's cultural background in a hierarchical or egalitarian society and the status or rank difference with its partner that an agent experiences against this cultural background. The following subsections specify the expected behavior with respect to these issues for each trade process.

Negotiation Behavior. Traders from small power distance cultures may have different ways to negotiate, but they will always negotiate. Traders from large power distance cultures on the other hand are not used to negotiating seriously. The powerful dictate the conditions. The less powerful have to accept. In cultures of large power distance that are also feminine or collectivistic the powerful may exercise restraint, or the lower ranked may successfully plead for compassion, but this is not a common decision making process, like a negotiation. The most powerful decides. When people from hierarchical cultures are forced to negotiate, because they are in a position of equal status or trade with foreigners, the negotiations often end in a game of power.

A trader from a culture with large power distance expects a lower ranked business partner to accept his conditions rapidly. If the lower ranked partner has the same cultural background, there is no problem and the rights of the higher ranked partner will be recognized and respected: the lower ranked opponent will be modest and give in easily. However, a trader from an egalitarian culture will not give in to the pressure if his status is lower, but will either react furiously (e.g., break off negotiations) or simply ignore the pressure (make a counterproposal), in which case the opponent will be furious (and e.g., break off negotiations).

If a trader from a culture with large power distance negotiates with a foreigner and assumes the foreigner to have a higher status, he may give in more easily than the foreigner expected. In that case the foreigner will be happy, but his opponent will have "left money on the table". If both are from hierarchical cultures but do not perceive one another's hierarchical position they may make misattributions resulting in one of them being dominated or stopping the negotiations.

Trade Goal Selection. Traders having a cultural background of small power distance opportunistically trade both low and high quality commodities and have a risk attitude that is not particularly influenced by power distance.

In hierarchical societies there are differences in selected trade strategy. The higher ranked prefer to trade high quality valuable commodities to underline their status that fits their position in life. They will not avoid deals where less powerful opponents technically have the opportunity to defect, because the higher ranked rely on their power to enforce cooperation.

The lower ranked have three incentives to prefer trade in low quality commodities in hierarchical societies. First, they know their place. Second, they are poor. Third, they may be cheated by high status opponents that make improper use of their power when trading valuable commodities; the lower ranked can avoid the risk of being deceived by trading commodities that have little appeal for higher ranked.

Maintenance of Beliefs about Partners. If counterparts have equal status, like in egalitarian societies, the experience of previous deals counts, be it personal experience or the experience of others (reputation). Failed negotiations decrease partner's future acceptability, and negotiations resulting in an agreement increase it for egalitarian traders.

In case of status difference in hierarchical societies, the acceptability of trade partners does not depend on experience in previous deals: the lower ranked have no choice but to accept business conditions and to show truthful and trusting behavior, whatever experience they have and whatever the reputations of their opponents are. So a lower ranked trader may have a belief about the fairness and trustworthiness of a higher ranked trader, but cannot show distrust. However, he may avoid a powerful trader that he believes to be unfair or untrustworthy... In a hierarchical society a trader of lower status forced into an adverse agreement by one of higher status will of course not find the opponent more acceptable afterwards, and try harder to avoid the opponent.

This cultural scripting may have its repercussions in intercultural trade. A lower ranked trader from a hierarchical culture might avoid a foreign trader if he assumed that his lower status did not allow him to negotiate successfully. A higher ranked trader from a hierarchical culture might overplay if he assumed that a foreigner would recognize his status and comply with his demands. An egalitarian trader who did not sufficiently respect the status of a hierarchical partner might fail to do business.

Truthful or Untruthful Delivery. After an agreement has been reached, it comes to delivery. If the quality of the commodity is invisible at first sight, the supplier can be opportunistic and deliver a lower quality product than agreed upon, thus making an extra profit. By doing so, regardless of the society's power distance, the supplier takes the risk of serious damage to the relation with the customer if the deceit be revealed.

For egalitarian traders, decisions to deceive, trust, and forgive are not influenced by their partner's status. Instead they depend on the quality of the relations they want to maintain and the sanctions they may expect.., In hierarchical societies, the higher ranked do not have to fear for repercussions when trading with lower ranked opponents, so the decision whether to defect or not merely depends on their morality, relationship, personality and/or circumstances. The lower ranked on the other hand will not easily consider to defect and will usually comply when trading with higher ranked and will only defect if in need. In collectivist societies, they would expect the higher ranked to recognize their need and to mercifully condone their behavior.

Trust or Trace. After delivery, the buyer may either request the delivered commodity to be traced, or accept it trustingly without tracing,.. In societies with large power distance, the lower ranked have no choice but to show trust in the higher ranked, whatever belief about their trustworthiness they may have. The higher ranked have no reason to distrust the lower ranked, because they assume that deceit of a higher ranked would not even be considered. So for the decision to trust, the belief about partner's trustworthiness is only relevant among equally-ranked or in relations where egalitarian traders are involved.

In intercultural contacts, the behavior of traders from hierarchical societies may be credulous in the eyes of their egalitarian opponents, because the high ranked rely on their status and the low ranked think it is improper to trace, thus encouraging deceit by the foreigners and eventually damaging the relation if deceit be revealed.

In egalitarian societies trust is equally important in every relation, regardless of partner's status. In these societies, decisions to trust a delivery or to request a trace (thus showing distrust) are not influenced by status difference. However, showing distrust may be harmful for relations, so there may be other incentives for benevolent behavior, but those incentives are not related to the dimension of power distance.

We assume that trust between parties will develop if negotiations succeed, even in the absence of positive evidence for the truthfulness of deliveries. On the other hand, a tracing report that reveals deceit will reduce trust.

Partner Selection. A trader has to select partners to deal with, either through response to a proposal made by another trader, or by proposing to another trader. Traders may use different criteria to select partners for new deals, according to their personal preferences and societal rules. The important criterion that differentiates partner selection across the power distance dimension is *status*.

In hierarchical societies traders will try to avoid partners who have higher status than they have themselves, because the higher ranked have the power to dictate business conditions. Traders will never propose business to higher ranked others because they are afraid of getting a bad deal. However, if they receive a proposal from a higher ranked trader they have to accept and the only thing they can do is plead and hope for magnanimity. Although one can exploit status in trading with less powerful counterparts, powerful traders in hierarchic societies prefer to do business with partners of their own level of power, because it would lower their status to get involved with people below their own standing.

In egalitarian societies status plays no role in partner selection. There are people who are labeled to have a high or low status in some respect like (show) business, politics, or sports. In strictly egalitarian societies, this will not influence the behavior of business partners. However in intercultural contacts, the traders from hierarchical societies may be influenced by the status labels of their egalitarian partners.

3   Representation in Agents

This section formalizes the knowledge about the influence of power distance on trade processes that was introduced in section 2. The relevant attributes of transactions from this viewpoint are the economic value of the transaction, the quality of the traded goods as a status attribute in its own ("we deal in superior quality products only") and a perceived risk that the trade partner will not fulfill his or her contractual obligations. The latter is based on trust in the supplier and attributes of the transaction, including product quality: highly valued products such as organic food, designer clothes, and jewelry are a more likely target for swindle and counterfeiting than are commodities. The formalization is based on DESIRE [11], an agent specification language based on information type definitions, process composition, and production rules.

The negotiation process is simulated using the negotiation architecture of Jonker and Treur [12]. The architecture is based on utility functions for comparing bids and a set of decision parameters. In this case we use the following utility function. In other cases, other types of functions may be appropriate, possibly involving additional attributes. In such case, some of the rules given later in this section may have to be adapted.

$$U_{\text{bid}} = w_1 f_1(\text{value}_{\text{bid}}) + w_2 f_2(\text{quality}_{\text{bid}}) + w_3 f_3(\text{risk}_{\text{bid}}) \qquad (1)$$

with $w_1 + w_2 + w_3 = 1$, and w_i in $[0, 1]$, for all i. f_1 presents the economic value of the bid in the interval $[0, 1]$; f_2 presents the additional value in $[0, 1]$ that is attached in society to trading in high quality products; f_3 evaluates the risk of swindle of the transaction in $[0, 1]$, with 1 representing a transaction without any risk.

Weight factors $<w_1, w_2, w_3>$ characterize an agent's trade strategy, e.g., *<high, high, low>* represent an *opportunistic* strategy, *<low, high, high>* a *quality-minded* strategy, and *<high, low, high>* represent a *thrifty* strategy.

Traders in extremely egalitarian societies do not adapt their trade strategy to partner's status. Traders in hierarchic societies do. Lower ranked traders in hierarchical societies prefer a more thrifty strategy than the higher ranked ones. The higher ranked follow an opportunistic or quality-minded strategy, depending on status difference with their trade partner. Let the relation agent_trait_value: ISSUE × Real, stand for the natural inclination of the agent to weigh an issue. Then the effect of the power-distance and the status of both parties can be implemented as follows.

```
/* 1 calculate weight factors using PDI and status */
if cultural_script_contains(power_distance_index(H: Real))
    and agent_label(status, S: Real)
    and current_partner(C: Trader)
    and partner_model_contains_belief(C, status, Y: Real)
    and agent_trait_value(value_preference, P: Real)
    and agent_trait_value(quality_preference, Q: Real)
    and agent_trait_value(risk_aversion, R: Real)
    and N: Real = P + (1-H)*Q+H*S*Q + (1-H)*R+H*(1-S+Y)*R
then utility_weight_for_value( P / N )
    and utility_weight_for_quality( ( (1-H)*Q + H*S*Q ) / N )
    and utility_weight_for_risk( ( (1-H)*R + H*(1-S+Y)*R ) / N );
```

Traits, status, and power distance index are real numbers in $[0, 1]$. This rule represents that - in proportion with the power distance index - the weight that traders attribute to trading valuable high quality products relatively increases with their social status, while the weight they attribute to risk relatively decreases with increasing feeling of superiority to the partner or increases with decreasing feeling of inferiority. The divisions by N normalize the sum of the weight factors to 1, so the weight of the economic value is indirectly affected by changes of quality and risk weights.

After evaluating a partner's bid with respect to value, quality, and risk, an agent has to decide whether to accept or to refuse the bid, and, in the latter case, whether to break off the negotiation or to make a counteroffer. Decision parameters are *utility gap* (difference of utilities that an agent will accept between partners' and own bid), *impatience* (probability that an agent will quit if utility or progress is low), *concession factor* (maximal relative concession with respect to the opening bid), and *negotiation speed* (maximal relative step toward maximal concession in a negation round). Furthermore, the rules use a *cut-off value* (minimal utility of partner's bid for which an agent continues) and a *minimal progress value* (minimal relative improvement of utility of partner's bids required in three rounds) as criteria to break off negotiations.

In the architecture of Jonker and Treur, agents accept an offer if the difference between partner's bid and their own bid is smaller than the *utility gap* parameter. As explained in section 2, negotiation in a hierarchical society is a game of power. The more powerful dictate the conditions of the deal. An agent from a hierarchical society feels forced to accept a bid of a more powerful partner even if the utility gap is not covered: the agent is aware that the utility of the bid would be unacceptable if it were made by a less powerful agent, but accepts.

```
/* 2 hierarchic agents accept sooner if partner is powerful */
if cultural_script_contains(power_distance_index(H: Real))
    and agent_label(status, S: Real)
    and current_round(X: Integer)
    and current_negotiation(C: Trader, X, L: Commodity_list)
    and partner_model_contains_belief(C, status, Y: Real)
    and agent_trait_value(acceptable_utility_gap, G: Real)
    and others_bid_utility_in_round(U: Real, X)
    and my_bid_utility_in_round(V: Real, X)
    and V-U < G * (1 + H * max(0, (Y-S)) * X )
then stop_negotiation(C, X, L, accept_offer);
```

Rules 3 and 4 express that in a hierarchic society an impatient agent will less likely break off negotiations with a more powerful opponent (suppressed impatience).

```
/* 3 have patience if powerful partners make unrealistic bids */
if cultural_script_contains(power_distance_index(H: Real))
    and agent_label(status, S: Real)
    and current_round(X: Integer)
    and current_negotiation(C: Trader, X, L: Commodity_list)
    and partner_model_contains_belief(C, status, Y: Real)
    and agent_trait_value(cut_off_value, M: Real)
    and others_bid_utility_in_round(U: Real, X: Integer)
    and U < M
    and agent_trait_value(impatience, I: Real)
    and random(0, 1, Z: Real)
    and I * (1-H*max(0, Y-S) ) * 0.5 > Z
then stop_negotiation(C , X , L, gap);
```

```
/* 4 have patience if powerful partners make no concession */
if cultural_script_contains(power_distance_index(H: Real))
    and agent_label(status, S: Real)
    and current_round(X: Integer)
    and X > 3
    and current_negotiation(C: Trader, X, L: Commodity_list)
    and partner_model_contains_belief(C, status, Y: Real)
    and agent_trait_value(minimal_progress, M: Real)
    and progress_in_bids(X-3, X, P: Real)
    and P < M
    and agent_trait_value(impatience, I: Real)
    and random(0, 1, Z: Real)
    and I * ( 1-H*max(0, Y-S) ) * 0.5 > Z
then stop_negotiation(C, X, L, no-accom);
```

Rule 2 is about accepting partners' bids. A hierarchic agent also accommodates a more powerful partner by making greater concessions in his own bids. In the architecture of Jonker and Treur: decrease the *minimum utility* parameter (rule 5).

```
/* 5 hierarchic agents give in easily if partner is powerful */
if cultural_script_contains(power_distance_index(H: Real))
    and agent_label(status, S: Real)
    and current_partner(C: Trader)
    and partner_model_contains_belief(C, status, Y: Real)
    and agent_trait_value(concession_factor, F: Real)
then   minimum_utility ((1-F)*(1-H*(0.5*(Y-S)+0.5*abs(Y-S))));
```

The *negotiation speed* parameter, i.e. the relative size of concessions toward the minimum utility, is not influenced by power distance. However, the absolute size of concessions increases with power distance, because concession size is the product of negotiation speed and the difference between the previous bid's utility and minimum utility.

The following rule (rule 6) is about the delivery, once a deal has been closed. Contracts may leave room for opportunistic behavior such as delivering goods of inferior quality. The decision whether to defect or to cooperate is modeled as comparing the temptation to deceive with a threshold (*honesty*, an agent's personal trait). The temptation depends on factors like the product quality agreed in the contract. In hierarchic societies the threshold for defection is influenced by status.

```
/* 6 hierarchic agents are conscientious with a powerful partner */
if cultural_script_contains(power_distance_index(H: Real))
    and agent_label(status, S: Real)
    and current_partner(C: Trader)
    and partner_model_contains_belief(C, status, Y: Real)
    and agent_trait_value(honesty, T: Real)
then   deceit_treshold(T+H*(Y-S)*(1-T));
```

The agents maintain a belief about the trustworthiness of other agents, i.e. the probability that they will not deceive. However, the decision to trust does not depend on this belief only. The relevance of this belief depends on two factors. First, the product quality agreed in the contract influences the relevance: expensive, high quality products are more sensitive to deceit than cheap, low quality products.

Second, in hierarchic societies the relevance of interpersonal trust for the decision to put deliveries to the test (trace) decreases as status difference increases (rule 7, which is only relevant for contracts about high quality product transactions). Low status agents do not trace high status agents, because they do not dare to show distrust. High status agents do not trace low status agents, because they trust that the opponents of lower status will not dare to defect.

```
/*  7 hierarchical agents do not trace if status difference is big */
if deal_in_round(C: Trader, B: Bid, X: Integer)
    and current_round(X: Integer)
    and cultural_script_contains(power_distance_index(H: Real))
    and agent_label(status, S: Real)
    and partner_model_contains_belief(C , status, Y: Real)
    and partner_model_contains_belief(C , trust, T: Real)
    and random (0, 1, Z: Real)
    and (1-H*abs(Y-S))*(1-T) > Z
then  to_be_traced(B);
```

Beliefs about partners are updated, based on experience. For trustworthiness belief:

$$t_{C,x} = (1-\delta^+)\, t_{C,x-1} + \delta^+\, e_{C,x}, \text{ if } e_{C,x} \geq t_{C,x-1},$$
$$t_{C,x} = (1-\delta^-)\, t_{C,x-1} + \delta^-\, e_{C,x}, \text{ if } e_{C,x} < t_{C,x-1}. \tag{2}$$

with $\delta^+ = \varepsilon\delta^-$ and δ^+, δ^-, and ε all in the interval $[0,1]$; $t_{C,x}$ represents trust in agent C after round x; $e_{C,x}$ represents the experienced result with C in round x. $e_{C,x}$ is either 1 (partner cooperated) or 0 (partner defected). Note that the model does not reason about the cause of the experience, e.g. by maintaining beliefs about partner's competence and honesty; the only thing that counts is the effect.

A similar update function is defined for benevolence. Being traced reduces the belief in partner's benevolence and not being traced is perceived as a confirmation of trust.

The belief about fairness is maintained similarly. For fairness the utility of the deal is used as experience value, a broken negotiation having an experience value of zero. When selecting partners, egalitarian agents compare others with respect to fairness. Hierarchic agents also use fairness, but their priority is to avoid status difference (rule 8). However, they cannot refuse if a higher-ranked proposes to do business (rule 9).

```
/* 8 hierarchic agents avoid partners with status difference */
if no_ongoing_negotiations
    and not_recently_proposed_to_me (C: Trader)
    and cultural_script_contains(power_distance_index(H: Real))
    and agent_label(status, S: Real)
    and partner_model_contains_belief(C , status, Y: Real)
    and partner_model_contains_belief(C , fair, F: Real)
then acceptability (C,  (1-H*abs(Y-S))*F);
```

```
/* 9 high-ranked partner is hard to refuse for a hierarchic agent */
if no_ongoing_negotiations
    and recently_proposed_to_me (C: Trader)
    and cultural_script_contains(power_distance_index(H: Real))
    and agent_label(status, S: Real)
```

```
      and partner_model_contains_belief(C, status, Y: Real)
      and partner_model_contains_belief(C, fair, F: Real)
    then acceptability (C, (1-H*abs(Y-S))*F + H*max(0, Y-S));
```

Details of partner selection that are not related to specific cultural dimensions are given in [1].

With respect to the decision making presented in this section the following must be noted. According to March [13], decision making can be modeled as either rational or rule-following. Equation (1) may suggest that the agents are modeled as rational utility maximizers. To some extent they are, as the first term of the function represents economic value of the transaction, e.g. the profit that a trader expects to gain based on market price beliefs and calculated risk. However, the other terms of equation (1) represent deviations from economic rationality that may be influenced by a trader's personality and culture. The second term represents an economically irrational preference for quality, for instance for dealing in luxury products in a situation where more profit can be made by dealing in standard products. The third term represents a risk aversion that goes beyond the calculated risk accounted for in the economic value of the transaction. Furthermore, the utility function is only used to valuate and compare bids during the negotiation process. All decisions about partner selection, accepting a bid, continuation of negotiation, deceit, and trust are modeled to be rule-following in the terminology of March.

4 Experimental Verification

The production rules formulated in the preceding section were verified in two steps. First, the formulation of the rules was verified in one-to-one agent scenarios. The rules were verified by step-by-step observation of the actions (duration of negotiations, quality levels and utilities of closed deals, break-off, trust and deceit decisions, and belief update) for different values of power distance index and status difference. Secondly, as DESIRE is not a suitable environment for simulating larger populations of agents, for verification of emerging properties at the macro level the agents were implemented in a multi-agent environment. For this purpose, CORMAS was chosen. CORMAS is a Smalltalk-based tool for multi-agent simulations that facilitates simulation with larger populations [14]. The verification results at macro level are discussed in this section.

Table 2 presents results of multi-agent simulations in single-culture and multicultural settings. Agents have the role of either supplier, customer, or tracing agent. In time step 1 the customers send a proposal to a supplier of their choice. In each next time step the trading agents may wait for a reply when they did send a proposal in the previous time step, or they may either reply with an acceptance message or a counter proposal, or ignore received proposals and take the initiative to send a new proposal to a preferred potential partner. If a deal has been closed, the supplier delivers and the customer may accept the delivery or forward it (at the cost of a fee) to the tracing agent. The tracing agent tests the quality, returns the delivery and reports its findings to the customer and the supplier. In case of deceit the tracing agent fines the supplier.

Table 2. Number of successful transactions in runs of 100 time steps. The agents are divided into two groups of four suppliers (S1 and S2) and two groups of four customers (C1 and C2). All agents have equal parameter settings, except power distance and status that may differ across groups. H stands for hierarchic cultural background (power distance index = 0.99); E for egalitarian (p.d.i. = 0.01); S for superior status (status = 0.8); I for inferior status (status = 0.2).

	run#	S1	S2	run#	S1	S2	run#	S1	S2	run#	S1	S2
run#	1	HS	HS	2	HI	HI	3	HS	HI	4	ES	EI
C1	HS	11	12	HI	8	10	HS	36	1	ES	17	13
C2	HS	19	13	HI	12	10	HI	0	23	EI	10	13
run#	5	ES	EI	6	HS	HI	7	EI	EI	8	ES	ES
C1	HS	33	0	ES	20	11	HS	10	14	HS	14	16
C2	HI	0	30	EI	5	13	HI	9	14	HI	16	17
run#	9	ES	EI	10	HS	HS	11	ES	EI	12	HI	HI
C1	HS	23	0	ES	20	13	HI	1	21	ES	10	11
C2	HS	26	0	EI	10	10	HI	1	24	EI	14	11

The results illustrate that the ease of trade depends on trader's status in hierarchic societies (runs 1 and 2). Trade stratifies according to status in hierarchic societies, but not in egalitarian ones (3, 4). In mixed settings stratification occurs especially when the hierarchic make the first proposal (5). Stratification is reduced when egalitarians make the first proposal, especially for the lower classes (6). When hierarchic traders have no choice but to trade with egalitarians they do so (7, 8). However, when given the choice they prefer peers (9-12). These results demonstrate that realistic tendencies emerge from interactions of agents following the rules specified in this paper.

5 Conclusion

There is a wealth of literature on trade and culture that so far has not been considered in formalized models of trade. In agent-based economics, individual traders are modeled as intelligent agents cooperating in an artificial trade environment. The agents are modeled to mimic authentic human behavior as closely as possible. In recent papers the differences between such agents is no longer solely attributed to differences in their individual economic situations. Aspects such as personality and attitude are considered as well, see for example, [15]. Without considering such aspects, the simulations will not correspond to reality. With respect to formalizing the important influence of cultural background on trade, we only found a few papers. These papers study trade at the macro-level. An example is [16]. This paper presents an equilibrium analysis on the amount countries invest in learning another language and culture and the size and welfare of those countries. Another example is [17]. That paper presents a formal model of the influence of trade on culture, i.e., the reverse direction of influence as studied in the current paper. Other literature also uses macro-level models, such as the gravity model to study the correlation between culture and trade, e.g., [18].

Most agent models of culture that can be found in the literature aim to adapt system behavior and user interfaces to the user's culture. Kersten [19] urges the necessity of cultural adaptation of e-Business systems and proposes an architecture that adapt both business logic and user interface. The rationale for adapting systems to user's cultures is given by Kersten et al. [4], who report significant differences in expectations, perception of the opponent, negotiation process, and outcomes of electronic negotiations across cultures. However, no actual implementations of models of culture in e-Business have been found to be reported. Blanchard and Frasson [20] and Razaki et al. [21] report an application of Hofstede's dimensions in a model to adapt e-Learning systems to the user's culture. Recent research on cultural modeling in agents mostly focus on Embodied Conversational Agents (ECA), including non-verbal behavior like facial expressions, gestures, posture, gazing, and silence in conversations; see, e.g., [22, 23]. For instance, the CUBE-G approach of Rehm et al. [24, 25] is based on the Hofstede dimensions and focuses on modeling into virtual characters the processes of first meeting, negotiation, and interaction in case of status difference.

All models discussed so far have in common that they model culture with the purpose to support human decision making or to improve human-computer interaction. The purpose of the model proposed in the present paper is to realistically simulate emergent behavior in multi-agent based simulations for research in the social sciences. The aspects of ECA are of less relevance in this context. Agent behavior may be modeled in a more stylized way. An approach that does so for the purpose of multi-agent simulations is that of Silverman et al. [26, 27]. They model agents as a composition of biological, personal (personality, culture, emotions), social (relations, trust), and cognitive (decision) modules, completed with modules for perception, memory, and expression. Their approach is a generic structure for modeling the influence of culture on agent behavior – along with factors like stress, emotion, trust, and personality – through Performance Moderator Functions (PMF). It differs from our approach in that it is an environment to implement validated models of culturally differentiated behavior, while our approach aims to develop and validate such models.

The contribution of this paper is the formalization of culture with respect to the influence of the power distance dimension on trade. This formalization has been carried out at the micro-level, i.e. at the level of individuals participating in trade. The traders' behavior is formalized in the form of rules that take power distance and status difference into account. The agents reason with a perceived model of the parties they consider for trading. These perceived models do not contain estimates of the culture of the other parties. Furthermore, the rules do not model the motivations and emotions that underlie the behavior. However, for study of macro-level effects as a consequence of cultural differences in micro-level interaction it is sufficient that the rules realistically model the effects of culture on individual behavior.

The work of Hofstede [8] offers detailed information to model the effect of culture on human behavior. Hofstede's model is based on a thorough statistical analysis of a massive amount of data. The five dimensions of culture discovered by factor analysis of the data are an efficient instrument to type national cultures. For each of the dimensions Hofstede's work offers extensively validated descriptions of difference in behavior along the dimensions. These descriptions are very well applicable to modeling differences in behavior of artificial agents.

The approach taken in this paper is to model a single one of Hofstede's dimensions. The other dimensions are treated in other work [1, 28, 29, 30]. Modeling a single dimension of culture is artificial. In reality, all aspects of cultural background have their effect simultaneously. However, modeling behavioral differences for a single dimension offers the possibility develop narrative descriptions of hypothetical behavior as presented in section 2 of this paper, to implement this behavior into agents, and to verify if the aggregated effects correspond with the expected behavior on the basis of Hofstede's theory. The results of the simulations presented in section 3 demonstrate that realistic tendencies emerge from interactions of agents that follow the rules specified in this paper.

Future work aims to integrate models for the separate dimensions into agents with believable culturally differentiated behavior, and to validate and calibrate the integrated models in two ways. First, business, economics, and negotiation science literature offer hundreds of papers that describe and analyze differences between cultures and intercultural interactions. Second, gaming simulations like [2] can be used to validate agent model behavior in specific configurations.

Acknowledgement. The authors thank John Wolters for engineering the multi-agent simulation in CORMAS.

References

1. Hofstede, G.J., Jonker, C.M., Meijer, S., Verwaart, T.: Modeling Trade and Trust across Cultures. In: Stølen, K., Winsborough, W.H., Martinelli, F., Massacci, F. (eds.) iTrust 2006. LNCS, vol. 3986, pp. 120–134. Springer, Heidelberg (2006)
2. Meijer, S., Hofstede, G.J., Beers, G., Omta, S.W.F.: Trust and Tracing game: learning about transactions and embeddedness in a trade network. Production Planning and Control 17, 569–583 (2006)
3. Adair, W., Brett, J., Lempereur, A., Okumura, T., Shikhirev, P., Tinsley, C., Lytle, A.: Culture and Negotiation Strategy. Negotiation Journal 20, 87–111 (2004)
4. Kersten, G.E., Köszegi, S.T., Vetschera, R.: The Effects of Culture in Anonymous Negotiations: Experiment in Four Countries. In: Proceedings of the 35th HICSS, pp. 418–427 (2002)
5. Gorobets, A., Nooteboom, B.: Agent Based modeling of Trust Between Firms in Markets. In: Bruun, C. (ed.) Advances in Artificial Economics. LNEMS, vol. 584, pp. 121–132. Springer, Heidelberg (2006)
6. Triandis, H.C., et al.: Culture and Deception in Business Negotiations: A Multilevel Analysis. International Journal of Cross Cultural Management 1, 73–90 (2001)
7. Tesfatsion, L., Judd, K.L.: Handbook of Computational Economics Agent-based Computational Economics, vol. 2. North-Holland, Amsterdam (2006)
8. Hofstede, G.: Culture's Consequence, 2nd edn. Sage Publications, Thousand Oaks (2001)
9. Hofstede, G., Hofstede, G.J.: Cultures and Organizations: Software of the Mind, 3rd Millennium edn. McGraw-Hill, New York (2005)
10. Hofstede, G., McCrae, R.R.: Personality and Culture Revisited: Linking Traits and Dimensions of Culture. Cross-Cultural Research 38, 52–88 (2004)

11. Brazier, F.M.T., Jonker, C.M., Treur, J.: Principles of Component-Based Design of Intelligent Agents. Data and Knowledge Engineering 41, 1–28 (2002)
12. Jonker, C.M., Treur, J.: An Agent Architecture for Multi-Attribute Negotiation. In: Nebel, N. (ed.) Proceedings of the Seventeenth International Joint Conference on Artificial Intelligence, IJCAI 2001, Seattle, Washington, USA, August 4-10, 2001, pp. 1195–2001. Morgan Kaufmann, San Francisco (2001)
13. March, J.G.: A Primer on Decision Making: How Decisions Happen. Free Press (1994)
14. Bousquet, F., Bakam, I., Proton, H., Le Page, C.: Cormas: Common-Pool Resources and Multi-agent Systems. In: Mira, J., Moonis, A., de Pobil, A.P. (eds.) IEA/AIE 1998. LNCS, vol. 1416, pp. 826–838. Springer, Heidelberg (1998)
15. Jager, W., Mosler, H.J.: Simulating human behavior for understanding and managing environmental dilemmas. Journal of Social Issues 63(1), 97–116 (2007)
16. Kónya, I.: Modeling Cultural Barriers in International Trade. Review of International Economics 14(3), 494–507 (2006)
17. Bala, V., Long, N.V.: International trade and cultural diversity with preference selection. European Journal of Political Economy 21(1), 143–162 (2005)
18. Guo, R.: How culture influences foreign trade: evidence from the U.S. and China. Journal of Socio-Economics 33, 785–812 (2004)
19. Kersten, G.E.: Do E-business Systems Have Culture And Should They Have One? In: Proceedings of the 10th European Conference on Information Systems, Information Systems and the Future of the Digital Economy, ECIS 2002, Gdansk, Poland, June 6-8 (2002)
20. Blanchard, E.G.M., Frasson, C.: Making Intelligent Tutoring Systems Culturally Aware: The Use of Hofstede's Cultural Dimensions. In: Proceedings of the 2005 International Conference on Artificial Intelligence, ICAI 2005, Las Vegas, Nevada, USA, June 27-30, 2005, vol. 2, pp. 644–649 (2005)
21. Razaki, R., Blanchard, E.G.M., Frasson, C.: On the Definition and Management of Cultural Groups of e-Learners. In: Ikeda, M., Ashley, K.D., Chan, T.-W. (eds.) ITS 2006. LNCS, vol. 4053, pp. 804–807. Springer, Heidelberg (2006)
22. Payr, S., Trappl, R.: Agent culture; Human-Agent Interaction in a Multicultural World. Lawrence Erlbaum Associates, Mahwah (2004)
23. Rehm, M., André, E., Nakano, Y.I., Nishida, T.: Enculturating conversational interfaces by socio-cultural aspects of communication. In: Proceedings of the 2008 International Conference on Intelligent User Interfaces, Gran Canaria, Canary Islands, Spain, January 13-16 (2008)
24. Rehm, M., André, E., Bee, N., Endrass, B., Wissner, M., Nakano, Y.I., Nishida, T., Huang, H.-H.: The CUBE-G approach – Coaching culture-specific nonverbal behavior by virtual agents. In: Proceedings of the 38th Conference of the International Simulation and Gaming Association (ISAGA), Nijmegen (2007)
25. Rehm, M., Nishida, T., André, E., Nakano, Y.I.: Culture-Specific First Meeting Encounters between Virtual Agents. In: Prendinger, H., Lester, J.C., Ishizuka, M. (eds.) IVA 2008. LNCS, vol. 5208, pp. 223–236. Springer, Heidelberg (2008)
26. Silverman, B.G., Johns, M., Cornwell, J., O'Brien, K.: Human Behavior Models for Agents in Simulators and Games: Part I: Enabling Science with PMFserv. Presence 15(2), 139–162 (2006)
27. Silverman, B.G., Bharathy, G., Johns, M., Eidelson, R.J., Smith, T.E., Nye, B.: Sociocultural Games for Training and Analysis. IEEE Transactions on Systems, Man, and Cybernetics, Part A 37(6), 1113–1130 (2007)

28. Hofstede, G.J., Jonker, C.M., Verwaart, T.: Modeling Culture in Trade: Uncertainty Avoidance. In: Proceedings of the 2008 Agent-Directed Simulation Symposium (ADS 2008). SCS, San Diego (2008)
29. Hofstede, G.J., Jonker, C.M., Verwaart, T.: Individualism and Collectivism in Trade Agents. In: Nguyen, N.T., Borzemski, L., Grzech, A., Ali, M. (eds.) IEA/AIE 2008. LNCS, vol. 5027, pp. 492–501. Springer, Heidelberg (2008)
30. Hofstede, G.J., Jonker, C.M., Verwaart, T.: Long-term Orientation in Trade. In: Schredelseker, K., Hauser, F. (eds.) Complexity and Artificial Markets. LNEMS, vol. 614, pp. 107–118. Springer, Heidelberg (2008)

Intrusion of Agent-Based Social Simulation in Economic Theory

Bogdan Werth and Scott Moss

Manchester Metropolitan University Business School, Centre for Policy Modelling,
Aytoun Street, Aytoun Building, M1 3GH Manchester, United Kingdom
{bogdan,scott}@cfpm.org

Abstract. This paper discusses the results of the agent based model's cross-validation with domain experts. Subsequently, the next research step – the conceptual design of the evidence based outsourcing model – is introduced and implementation difficulties are discussed. Both models are developed in the course of the ongoing PhD research on outsourcing behaviour at financial institutions and are validated and discussed in this context.

Keywords: agent-based modeling, outsourcing, offshoring, transaction cost theory, transaction cost economics, investigation of economic theory, model validation.

1 Introduction

Newspaper headlines this summer were dominated by the turmoil of the credit markets. Reasons for this were, inter alia, the unexpected rise in US subprime mortgage delinquencies, the resulting surprisingly big losses reported by banks and eventually the sharp shortage in liquidity. The aftershock phase is now characterized by deeper recession. The course of these events did not help the banking sector to establish a reputation for having reliable risk management systems.

One possible reason for the banking disaster can be found in the drain of appropriate knowledge assets necessary for crisis management. "The effects of the subprime crisis on our business are better than one would assume initially since banks depend on us to audit their risk management systems. During the whole outsourcing hype of the early 90s banks lost many qualified IT personnel and therefore could not retain the necessary know-how to cope with their problems"[1] so the director of the IT-efficiency department at the Financial Services Advisory practice in PricewaterhouseCoopers Germany. This is by far, not an exclusive opinion. The Chairmen and President of the Frankfurt School of Finance & Management Udo Steffens suggests in [1] that the misery of the banking sector originates from the fact that the involved institutions simply did not have enough experience in order to assess the real risks.

As studies show, the companies that are extremely successful over a substantial period of time are characterised by the ability to carry out radical strategic changes without imperilling the operational effectiveness.[2] The question of what services

---

[1] Author's own translation of the original citation from German language.

N. David and J.S. Sichmann (Eds.): MABS 2008, LNAI 5269, pp. 17–32, 2009.

and functions a company should produce in-house versus what it should acquire from third party vendors is at the heart of this process.

The rapid economic growth of the late 1990s and the sudden crash of the early 2000s reinforced the wave of re-engineering issues in many corporations. During the downturn, outsourcing leaped to the fore as a cost-saving quick fix when budgets were squeezed. The cult of core competency cultivated by Prahald [3] and Hammer [4] led to a decade of re-engineering and downsizing. One theory which ties up with this cult in the IT/S sourcing research is the Transaction Cost Theory (TCT) [5].

The transaction cost theory (TCT) model [6] reported here was used to investigate a particular issue of the TCT concerning the formalization of the term asset specificity. The results of the cross-validation of the model through domain experts are reported. While the results were in line with the theoretical literature, they diverged significantly from the experience of our informants.

This paper discusses the results of the model's cross-validation with corresponding domain experts. Subsequently, the conceptual design of the next logical research step – an evidence based outsourcing model (EBOM) – is introduced and implementation difficulties are discussed. Both models, the TCT and the EBOM, are developed in the course of the ongoing PhD research on outsourcing behaviour at financial institutions.

2 Methodology

A methodology of evidence-based modelling was adapted. The rules for the agents were derived partly from the relevant reports and partly from qualitative insight into the modelled target system. These insights were gained from semi-structured interviews with domain experts. Internal and published support documents were collected.

Since it is common practice in sourcing projects for the analysis to be provided by third party consultants, these were selected as a target group of stakeholders. It should be remembered that the sample used for qualitative studies was opportunistic.

The TCT model includes a direct representation of relevant relations from the transaction cost theory. The main subjective of the presented model is the reproduction of the individuals' behaviour inspired by TCT and willingly abstract from any thoughts of social components that might matter in the target system. However, the original theory needed to be altered in a way that made it possible to translate the theory in an agent based model. Since Williamson's framework belongs to the realm of equilibrium theories and can be regarded as a snapshot in time, it did not allow for translation of the processes.

Before the conceptualization of the EBOM a cross-validation round with stakeholders was scheduled. The ulterior motive for this is the reinforcement of the feedback loop in the design process of the model. The TCT model is a formalization of the economic theory with evidence based elements in it. Since the used theory is controversial in the academic world but, nevertheless, widely used by practitioners, the results of stakeholder's feedback were hoped to be enlightening. Through the process of cross-validation there was evidence of shortcomings of the theory. These are valuable contributions for the next research round and also for the awareness of stakeholders. Next chapter brings out the gist of the validation process with industry collaborators.

3   Stakeholder Validation

In the following chapters the results of the model validation in cooperation with domain experts are reported. The stakeholders were presented with the model setup, a representative run and the results of the parameter exploration. The issue of detail is pragmatic. The level of detail chosen should enlighten the relationships of interest and hide unwanted detail and was previously discussed in earlier interviews.

3.1   (Production) Space

The n-dimensional product characteristics space (PCS) [7], similar to the approach taken by Klos in [8], was used in an attempt to represent the core TCT concept of "asset specificity". The Lancastrian PCS approach was adopted for the TCT model in such an altered way that instead of products and consumers, the space has now been populated with agents representing vendors and consumers. This way, one could define product differentiation and asset specificity on the same space. Thus, the location of the vendor agents represents the service he offers and the location of the client agent represents the in-house service he would like to outsource.

Thus, with the help of this apparatus one can define the specificity of own services by the number of agents who can provide the same service. The degree of specificity evaluation is then a subjective function of the corresponding agent.

This interpretation of asset specificity was welcomed by the stakeholders since it reflects the way companies evaluate standardisation of their IT services – if someone else on the market can provide the needed service without extensive alterations of the company's workflow then the service is deemed to be standard.

Another virtue of PCS is the possibility to express the concept of bounded rationality through the acceptance range of each agent. This construct was in line with the essence of bounded rationality – that agents' computational and information processing capacities are limited [9] – and correspond with the experiences of stakeholders – that depending on the size of the company, it will not be possible to survey all appropriate vendors on the market.

Furthermore, Lancaster's approach will allow for interface to extend the results of the EBO model to consumer theory. Investigating sourcing theories under the aspect of consumer theory is a promising, though uncommon for sourcing community, approach.

3.2   Opportunism[2]

Williamson's framework is interwoven with the notion of opportunism. This is the drive of every agent since TCT does not allow for mutual thoughts. Both agent types, client and vendor, re-evaluate their situation at every simulation step and are prepared to terminate the ongoing partnership in the case that an improvement of the agent's current situation is possible. The ulterior motive for this agent's behaviour is the notion of opportunism.

---

[2] The sort of opportunism described in this section deals with opportunistic behaviors, which may occur during a transaction between two agents.

However, this behaviour caused disagreement amongst most of the stakeholders interviewed. One does not disclaim the existence of opportunism on the executive floors of every corporation. Admittedly, the aforementioned opportunistic behaviour should not be such a pivotal part of a sourcing partnership. According to the domain expert from Accenture, in most outsourcing negotiations a vendor is set up as an enemy, which has to be forced into doing what the buyer of service thinks is best. Negotiation is usually set up around driving cost down, which is normal for the procurement activity. Notwithstanding, sourcing contracts are very different from normal procurement. It is a question of setting up a relationship which lasts for several years and cannot be treated as a one-off deal. "If we have a good social relationship with our sourcing partner, most of the time, it embodies in a good professional partnership. However, a dysfunctional social relationship holds lots of surprises which can rub off negatively on the professional level", so the procurement CEO at T-Systems.

3.3 Preference and Matching

The fact that any sourcing partnership is initiated by the client was endorsed. Thus, stakeholders pushed for a wider awareness level of clients about the existence of possible vendors. Correspondingly, proportions of the service providers which should be known to a client, with respect to the overall outsourcing market, were specified.

A disagreement was evoked by limited facets of the partnership search routine. In the current model a client would choose his contractual partner on the basis of the best match to his characteristics (i.e. vendor with shortest distance to the client on the PCS). In case a client cannot find any suitable vendor, even after altering his acceptance range, he will be matched to himself, thus preferring in-house production over outsourcing. The whole range of consideration aspects in the outsourcing process are discussed in further detail in paragraph 4.

3.4 Adaptation

Some vendors, who did not get any requests, will relocate randomly one step in any direction in the production space. This is an attempt to alter services' characteristics in order to look more appealing to the clients, which might not have considered this vendor otherwise. The customers, to the contrary, are assumed to be stationary and alter their acceptance range with respect to what is an acceptable service level agreement to them.

The domain experts suggested the consideration of cases where one would encounter joined sourcing i.e. situations where several small banks would adopt a common services portfolio in order to deliver enough scale for the vendor. Standardisation is not only an objective of the clients. The domain experts have mentioned that outsourcing companies try to standardise their portfolio of products in order to grow whilst keeping costs under control.

The willingness of the agents to terminate their ongoing relationships was questioned by stakeholders. Despite being in line with the TCT, termination was said to be an uncommon tactic in the experience of stakeholders. It was pointed out that due to legislative restrictions, most of the contracts will terminate in the natural way.

Indeed, some disaster scenarios are possible where an ongoing relationship can be terminated prematurely. The domain experts specified probabilities for such an event to take place.

3.5 Representative Run

The results from a representative run over 10.000 simulation steps with 1000 vendors and 500 clients were presented to the audience of domain experts.

In the rule-based system the number of activations can be used as an indicator for activity of the system. In figure 1 the cumulative number of rule activations for client/vendor agents is represented. In addition, the amount of ongoing transactions at each time step of the simulation is laid in the same chart. Correspondingly, in figure 2 the number of client-agents who ended up in the lock-in[3] is presented.

The diagram of activations in figure 1 exhibits that the model is constantly fluctuating, therefore agents are either applying topological or/and general reasoning rules[4]. The charts suggest that the system is in continuous dynamic though it has the tendency to gradually lose on activity. Both the number of rule-activations for the clients as the number of ongoing transactions in the system are decreasing rapidly at the beginning and slowly in the later stages of the simulation.

At the start of the simulation, a large number of rule activations are observable. This is partly due to the setup rules which fire at the beginning of the simulation but also partly due to the high degree of interaction between vendors and customers at the start of the simulation as they are seeking the contractual partners. The use of agents' memory economises on re-evaluations at a later stage of the simulation.

Domain experts were confronted with one possible explanation for the observed behaviour, namely, due to the ex ante negative experiences with some vendor-agents in the past, and the client-agent would not consider these vendor-agents as a viable alternative. Therefore the client-agent will remain in his current transaction despite being deeply dissatisfied, coupled with the availability of malicious vendors, lock-in is created. Furthermore, the selfish nature of the TCT prescribes the agents to break up the relationship as soon as they see a better alternative rather than their current transaction partner. Consequently, there is a high level of activity to observe during the earlier stages of the simulation – in the so called mating stage – where everybody is aspired to find the perfect match. Subsequently, the amount of possible alternatives (vendor-agents without a negative experience track record) for the client-agents decreases and so his activity thus bringing an overall sedation in the system.

Experts reported that figure 1 exhibits, approximately, the state of the sourcing market since the economic downturn of the early 2000 when outsourcing leapt to the fore as a cost-saving quick fix when budgets came under squeeze. There was a lot of turmoil on the sourcing market since everybody was outsourcing compulsively. Since then the market exhibits some activity but can be regarded as being in a more or less

---

[3] A client-agent is thought to be in the lock-in if he is not satisfied with the transaction he is currently involved in but cannot drop out of it (be it due to the lack of alternative vendors or personal preferences of the client-agent).

[4] The difference between the topological and reasoning rules as well as rules itself are described more in-depth in [6].

semi-stable state. However, the constant decrease of transactions and activity could not have been verified by all interviewed domain experts.

The explanation of the model behaviour the experts were confronted with was accepted. However, it was pointed out that, most of the time, these are not the negative ex ante experiences which deter banks from switching providers but the "patron" effect. It is cheaper and less risky to subcontract to someone you know for a long time instead of opting for a newcomer.

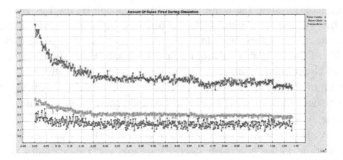

Fig. 1. Number of rule activations for all agent types and transactions. Upper line – clients' rules; middle line – transactions; lower line – vendors' rules.

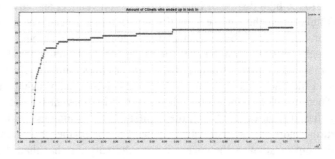

Fig. 2. Number of client-agents who ended up in the lock-in situation

3.6 Dynamic Equilibrium

In the forefront of results discussion with domain experts an objective of the equilibrium behaviour of the TCT model had high topicality. During parameter exploration an issue of the stable behaviour of the model was examined. In order to find out whether the system resides / converges into a sort of dynamic equilibrium, a perturbation run was conducted[5].

As one can clearly see on the presented charts in figure 3 and 4 there is a big jump in all the graphs subsequent to a new batch of agents being inserted into the system. The overall intensity of activations increases in value, nevertheless the typical shape

---

[5] For some time after the model run the number of clients and vendors were doubled. New agents were injected into the system and randomly distributed over the production space.

of the curves remains the same. New constellation produces the same long and stable periods of rule activations.

This behaviour was confirmed by domain experts. Undoubtedly, any market can exhibit volatile and unforeseen behaviour. However, sourcing market at the investigated target system – big financial institutions – featured equilibrium-like behaviour since it was introduced by IBM a few decades ago. In the face of the current turmoil in the banking industry, experts showed a great deal of interest in the factors which can bring such a system to be out of balance. This task remains to be answered by the EBOM.

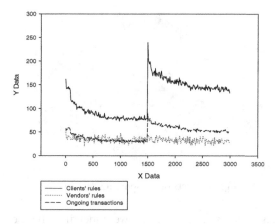

Fig. 3. Number of rule activations and ongoing transactions

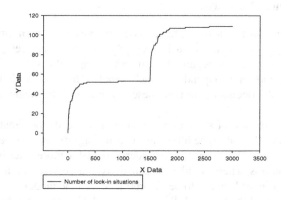

Fig. 4. Number of client-agents who ended up in the lock-in situation

4   Implication of Validation for EBOM

The discussed TCT model is an intermediate stage in the course of the ongoing research project on outsourcing in financial institutions. Therefore, the next logical step would be to use the feedback from the process of cross-validation for the development of the next stage model. This EBO model should be based entirely upon the evidence from the stakeholders and should not be constrained by any economic or behavioural theory.

Building the TCT model and confronting stakeholders with it was intended in order to point out data requirements, and help to determine which data is important and which can be dismissed. In the current research stage additional interviews were conducted in order to fill gaps, which became evident through the analysis of the TCT model. In the course of the cross-validation with stakeholders several issues arose, which are reported in detail in the following sections.

Fig. 5. Facets of the sourcing decision

Overall, six facets of the sourcing process were identified by the domain experts. Figure 5 illustrates these different aspects of the sourcing decision.

4.1 Co-dependence of Services

While outsourcing is a not a new phenomenon by any stretch of imagination, for much of its history it has been primarily focused on discrete services [10]. However, IT has evolved and cannot be regarded as an assemblage of discrete services. These services both support and depend on each other. The notion that services are autonomous and thus decisions around these services can be made autonomous is absolutely wrong.

As an increasing amount of services are getting outsourced, coordinating a portfolio of service providers is becoming tedious, thus causing significant service disruptions in numerous organizations. Therefore, stakeholders urged for more intertwined sourcing relationships. Scenarios where vendor A can deliver a service for a client only in collaboration with vendor B needs to be incorporated into the model.

The representation of the firm's skills universe (FSU), as was done by Tylor in [11], provides a possible way of dealing with the whole services agglomeration of the company. According to the experience of stakeholders and theories of Prahald [3], one would assume roots of the graph as core services thus all other services depending and predicating on these core services. Therefore, a dislocation of the service to the third party vendor can be expressed as an extraction of the corresponding node with its links from the FSU of the client.

4.2 Sourcing Partnership

During the validation process the awareness emerged that with respect to relationships one should define the following bank profiles – the risky ones and the risk averse ones – with the consequence of reflecting diverse outsourcing strategies. It is regarded as risky to streamlining a corporation to its essential elements with only one sourcing partner. A distribution of the service failure probabilities on multiple contracts is a safer option. Therefore, the agents with high risk aversion would preferably enter into the multisourcing[6] relationships rather than relying on a single vendor.

It was also pointed out that simple relationships such as those implemented in the TCT model are an extreme oversimplification. A more interconnected construction of partnerships with multiple sourcing partners, have to be implemented in the new model. "It is not enough if we have a good relationship with our vendors. Since they can deliver services we need only by working hand in hand with each other it is pivotal for them to have a good cooperation too" so the CFO of T-Systems. Thus a network topology of the contractual relationships will reflect the circumstances.

Furthermore, experts have specified three types of the sourcing deals: effectiveness, acquisition of new services and transformation & change. Each of the contract types has different setups, pre-requirements and outcomes.

Another artefact of the sourcing market is the capacity and scale savings of the vendor. One should not automatically assume that the service provider would be able to deliver economies of scale and so deliver the services I used to have at the discounted price. The only way the service providers can deliver an unprecedented level of service at a discounted price is through increasing scale and standardized products. Thus, the client has to create and govern the scale in a move towards a more standardised service portfolio.

4.3 Social Aspect of the Partnership

The view of technology in general as a tool which is considered to be independent of the social context, in which it is developed and used, is widely spread in the information systems literature [12]. The majority of domain experts agreed on the importance that information technology cannot and should not be regarded separately from the social context it is situated in. However, a precise statement with respect to this topic could not be achieved. Therefore, stakeholders were urged to investigate this issue with respect to the volatility of the equilibrium state of the system. It was reported about the projects which have failed due to social resistance of the company employees. The further discussion on this topic led to the strategy & change issue.

4.4 Strategy and Change

The main shortcoming of the TCT is the absence of strategic consideration possibilities. Williamson's framework does not allow for making any business strategy considerations. Also, TCT does not allow for translation of processes since it takes a snapshot view of the corporation.

---

[6] A new operational model that envisages provisioning of business services from multiple sources inside and outside the corporation to obtain the best business outcomes.

Absence of representation of business and outsourcing strategy was the biggest deficiency of the TCT model criticized by stakeholders. According to their experience the misalignment of business and outsourcing strategy scuppers the majority of sourcing projects. First, aligning the sourcing strategy with the business strategy of the company is pivotal for the achievement of the set goals. "If a company plans to enter into a sourcing relationship in order to expand, she cannot use a cost minimizing sourcing strategy", as stated an IT-consultant from Accenture. Second, if the goals of the sourcing project are not communicated well enough down the company hierarchy one will result in high personnel attrition rates and loss of tacit knowledge.

Furthermore, the adjustment of the business and sourcing strategy to each other has to be done on a continuous basis. A base line view of operations set up on the current knowledge cannot stay the same over the lifespan of a sourcing partnership since operations do not possess a steady state.

From an implementational point of view, the term strategy is difficult to grasp and translate into the model. Therefore, experts have agreed to the following definition – the enterprise's sourcing strategy is the bundle of planned actions for the achievement of set business goals. Thus one has two sets of items that have to be linked. By adopting this definition of the strategy, it is possible to fall back on earlier work of Moss [17]. In the critical incident management model agent's cognition is presented by problem space architectures – which are in effect relationships between goals and the sub-goals to achieve these goals – drawn from cognitive science. The construct is then rounded up by the use of Endorsements [13].

The drain of knowledge in the course of inadequately managed sourcing actions can easily be implemented by the use of FSU [11]. Thus an extraction of the corresponding node with its children from the skill graph represents loss of information. The closer the location of the extracted node to the root, the bigger the damage to the tacit knowledge of the corporation.

4.5 Asset Specificity

Formalization of Asset Specificity was one of the main objectives for the TCT model. An attempt to translate a loosely defined term into a formal simulation model can be regarded as successful since all of the interviewed stakeholders approved the suggested formalization – subjective function based on the amount of other agents able to provide considered service. However, an institution of asset specificity, or core competency as practitioners tend to call it, remains a term which is hard, if not impossible, to define precisely. After in-depth discussions, the majority of experts referred to intuition and own experience as a guidance for core competency evaluation. The words of the senior consultant in the IT practise group at BCG reflect the situation accurately "It is our biggest challenge to find out the thin line which separates what can be standardized and what has to be customised".

5 EBOM

This chapter introduces the design of the EBOM that is being developed under implications provided in the previous chapter. Whereas TCT model is build around a theory and matched to evidence in consequent steps, the EBOM follows a different approach – it is build around evidence first and is going to be investigated on a match to possible theoretic frameworks at a later stage. Thus the insights obtained after evaluation of the TCT model were constitutional for the EBOM's design.

5.1 The Setting

Similar to TCT model, a rather strict distinction between physical and social environment of agents was made. The focus of this distinction was put on separation between physical and social spaces, both in terms of semantics and techniques used for their representation. Furthermore, since agents are considered here in more than one social context an agent's social environment consists of and is modelled by more than one network layer.

A bird's eye view of the model is provided in figure 1. It showcases the interplay of different components composing the EBOM. The model represents actors from the case study as individual agents. Actors in this context are players of the outsourcing market – clients and vendors.

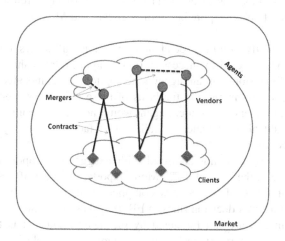

Fig. 6. Schematic view of the model's components and their relations to each other

The model is run at monthly steps. For each year (consisting of 12 ticks) the model executes the following sequence of events: agents proceed with re-evaluation of their situation, alter their strategies according to the results of their evaluation and act accordingly.

More precisely, clients aim at either reducing their costs or gaining new skills. Both objectives can be accomplished trough outsourcing, in case of cost reduction, and sourcing, in case of acquisition of new skills. Clients interchange their strategies depending on their current situation. In the process of agents' interactions clients are playing an active part, since they trigger the interactions. A client would broadcast a service request (SR) through the system and vendors, who can match the request, will respond with offers. Depending on an evaluation of SR respondents a client will either send a contractual offer or not. A vendor, in her turn, will perform an evaluation of all received offers based on several vendor-specific preferences with regard to clients. Once the deal is accomplished it is followed by the transition of skill sets from the vendor to the client. During a transaction each of the two partners may opt out of the contract prematurely or stay committed until the natural end of the transaction. In the case of a premature end of the transaction, sanctions are imposed on the defecting partner.

Vendors, among themselves, are able to enter into merger relationships. These can happen in a volunteer manner, thus two suppliers deciding to collaborate, or in a hostile manner, thus an underperforming supplier is getting absorbed buy a market rival.

5.2 Agents

Generalized agent types were determined based upon elicited knowledge, i.e. the storylines derived from the transcripts of the interviews. These agent types differ along dimensions like activities related to goal achievement and social network integration. Thus, the model is based on the definition of two different kinds of agents that perform different tasks. A population of N clients and M vendors with $M > N$ is assumed.

In order to explicitly contrast social and economic influences on the decision making of agents with regard to success of (out)sourcing relationships, two decision making driving dimensions of an agent's perception were introduced: economic success and social environment. On the basis of these two factors agents decide on their collaboration partners, i.e. a client agent would not only consult a financial side of a deal but would also evaluate, so called, soft factors (ethnicity of a vendor, vendor's reliability, etc.).

By purpose, agents were not endowed with exaggerated opportunistic drive *a priory*, as was done in the case of TCT. More importantly, a set of abstract decision rules for different types of actors was compiled, which forms the basis for the implementation of agent's decision rules in EBOM. These rules are entirely based on stakeholders' input and are not burdened by the theory. Thus, it is to investigate, whether opportunistic patters can emerge from the given set of rules.

5.3 The Network

In order to capture intrinsic properties of the modelled social system a multilayered network of agents was required. EBOM's environment represents agents' relationships as links in a network of multiple layers. An agent may be seen as a node in different social network contexts. Thus, an agent has two semantically different meanings depending on the context it is related to – sourcing relationships network (clients2vendors) and organizational relationships network (vendors2vendors).

Neither the merger nor the collaborator networks exist initially. Thus, these networks are created from the scratch each time simulation is run. Two networks differ in complexity depending on interactions of agents. The network is dynamic in a sense that clients and vendors are constantly establishing new relationships and terminating old ones.

5.4 FSU

During the validation of the TCT model it became evident that, majority of interviewed stakeholders used an internal IT representation that resembles strong similarity to a spanning tree structure. Nodes were referred to in diverse manners - hardware agglomeration points, knowledge or personnel skills of a corporation's unit, etc. Thus, for the EBOM nodes were interpreted as a collection of tasks or so called skill sets (SS) needed to get to the next node on the branch.

It is believed that a directed spanning tree is an adequate representation of interplay of goals with SSs necessary to achieve these goals. A similar concept was already used by Moss [17] and Taylor [11].

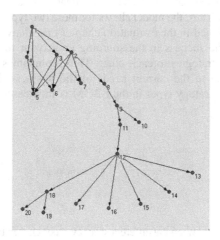

Fig. 7. Representative Firm Skills Univers of 20 nodes

From figure 7 it becomes clear that while some SSs strongly depended on several basic (parent) SSs, others, in contrary, developed more or less independently as branches of the tree.

An issue of asset specificity represented a *sine qua non* while implementing the model. Despite attempts to confine a focus of stakeholders on a standard set of "skills", the notion of an asset being more specific than others was omnipresent. EBOM introduces, besides other behavioural indices, a definition of asset specificity inspired by the initial findings of TCT model's development and evaluation process. TCTM deals with AS by estimating the number of providers, who can provide a particular service of interest. Considering, that this notion was endorsed by practitioners it was to determine now, how to incorporate the aforementioned way of going AS issue into EBOM.

A challenge arose from the fact that TCTM did not distinguish between different services in a client's portfolio but aggregated to a single service. Together with practitioners a notion of a redundant service was produced. Hence a redundant service could be interpreted as a service that exhibits alternative ways of delivering it. Putting this finding it into a relation with a spanning tree concept means, that any node, casted as redundant, has to be reachable form a root node via several paths. An in-degree could be used in here as a specificity indicator of a SS. A distance to the root node can be utilized as an indicator too – the closer is a SS to the root the more core-specific it is.

5.5 Strategy

Before an agent can enter a sourcing relationship it has to define its sourcing strategy. The TCT model's validation revealed that absence of strategic considerations was the biggest shortcoming of the model. It was hard to form an assemblage of theories from a plethora of strategic guidelines. However, according to the practitioners' statements all strategic considerations can be reduced to a two generic once – transformation/enhancement and efficiency. Efficiency deals are concern primarily with reducing costs, i.e. outsourcing in-house services and transformation/enhancements deals are concerned primarily with acquisition of new skills. Thus, the model allows for these two types of contracts.

It was already mentioned in the evaluation chapter that many stakeholders assigned a high proportion of the success in the sourcing endeavour to the alignment of the sourcing and business strategies to each other. Thus, companies switch between both strategy-types according to the current needs of the business. The dynamics of the interplay between both strategy types in the EBOM is showcased in the figure 8.

Fig. 8. The interplay between the different strategy types chosen by the client agent

Stakeholders urged for several strategies for clients, since the change between several strategies would represent the dynamic nature of business or agility in response to changing business circumstances.

6 Discussion

The example of the TCT showed that ABSS can be utilised as an appropriate tool for theory examination. Therefore, we need a more critical observation of theories borrowed from different disciplines before these can be established. The methodology presented in the paper aims to point out obscurities where the theory is not clear enough. As the process of formalisation is a discipline which needs to be precise, ABSS claims to compensate for the lack of precision in the IT/S research so far. ABSS uses models to devise precise statements about these theoretical points, which are not clear in the theory a priori.

The proposed formalisation of asset specificity used for agent's reasoning succeeded in reproducing results forecasted by TCT and offered a qualitative description of the term. Furthermore, the model and results produced by it were partly confirmed by domain experts used for cross-validation. Organisational structures emerged from the interaction and information exchanges between individuals in the artificial world. It was shown that organisational structures depend on connectivity and information exchange between agents.

Nonlinear social behaviour needs to be included in the outsourcing research on the more elaborate basis rather than just excluding it from the outset per asylum to economic theories. Nonlinear behaviour is excluded in most econometric observations as the concept of nonlinearity leads to problems for the statistician and their forecasts.

The necessity to model economic processes and theories used for decision making in the practise becomes more eminent if one considers the recent example of Société Générale. Who, from the involved bank staff, could have thought that a single trader could produce such extensive damage without being discovered by the technology systems, which have checks and balances built in? The French bank made a loss of £3.6bn [14] following unauthorised activity of a swindler trader who managed to cover up fraudulent activity as a result of his understanding of the bank's fraud control systems. He risked billions by betting on future trends in the stock market. The failure to detect fraudulent activity of this scale from an employee emphasizes the need for banks to constantly monitor and upgrade their fraud systems.

History is, however, being repeated. In February of 1995, one man single-handedly bankrupted the bank that financed, inter alia, the Napoleonic Wars, and was Queen Elizabeth's personal bank – Barings bank [15]. These incidents make the usefulness of agent based modelling in economic theory eminent.

While it has been addressed by many academics and practitioners in a variety of ways, it is still a challenge for most organisations to identify what to outsource and what to retain in-house. Moss and Edmonds showed [16] that appropriately specified models can be used to assist domain experts identify and amend dissonance between their own qualitative judgements and quantitative relationships drawn either from statistical or physical theory. This is the path the described outsourcing research is intending to follow in order to produce an agent-based model, which can highlight different aspects of, and provide valuable insights into the sourcing process.

References

1. Mohr, C.: Wie viele Jobs kostet es diesmal? Junge Karriere 03-08, pp. 48–50 (2008)
2. Moeller, M.: Return on Strategy. In: Boos, F., Heitger, B. (Hrsg.) Wertschöpfung im Unternehmen. Wie innovative interne Dienstleister die Wettbewerbsfähigkeit steigern, pp. 61–80. Gabler Verlag, Wiesbaden (2005)
3. Prahald, C.K., Hamel, G.: The Core Competence of the Corporation. Harvard Business Review (1990)
4. Hammer, M., Champy, J.: Reengineering the corporation. Harper Collins, New York (1993)
5. Aubert, A.B., Weber, R.: Transaction Cost Theory, the Resource-Based View, and Information Technology Sourcing Decisions: A Re-Examination of Lacity et Al.'s Findings. GreSI publications (May 2001)
6. Werth, B., Moss, S.: Transaction Cost Economics meets ABSS: a Different perspective on Asset Specificity in the IT-Outsourcing context. In: The Fourth Conference of the ESSA, Toulouse, France (2007)

7. Lancaster, K.: A New Approach to Consumer Theory. Journal of Political Economy 74, 132–157 (1966)
8. Klos, T., Nooteboom, B.: Agent-based computational transaction cost economics. Journal of Economic Dynamics and Control 25(3-4), 503–526 (2001)
9. Moss, S., Dixon, H.D., Wallis, S.: Evaluating Competitive Strategies (1994)
10. Cohen, L., Young, A.: Multisourcing - Moving Beyond Outsourcing to Achieve Growth and Agility. McGraw-Hill Professional, Harvard Business School Press (2006)
11. Taylor, R., Morone, P.: Modelling Knowldge Production and Integration in Working Environments. In: Edmonds, B., Iglesias, C.H., Troitzsch, K.G. (eds.) Social Simulation: Technologies, Advances and New Discoveries. Idea Group Publishing (in press)
12. Willmott, H., Bridgman, T.: Institutions and Technology: Frameworks for Understanding Organizational Change—The Case of a Major ICT Outsourcing Contract. The Journal of Applied Behavioral Science 42(1), 110–126 (2006)
13. Werth, B., Geller, A., Meyer, R.: He endorses me – He endorses me not – He endorses me. Contextualized reasoning in complex systems. In: Papers from the AAAI Fall Symposium: Emergent Agents and Socialities: Social and Organizational Aspects of Intelligence, AAAI Technical Report: FS-07-04, Washington, DC, November 8–11 (2007)
14. Handelsblatt: Schock Générale, January 28 (2008)
15. Bonabeau, E.: Predicting the Unpredictable. Harvard Business Review, Reprint R0203J (2002)
16. Moss, S., Edmonds, B.: A knowledge-based model of context-dependent attribute preferences for fast-moving consumer goods. Omega 25, 155–169 (1997)
17. Moss, S.: Critical Incident Management: An Empirically Derived Computational Model. Journal of Artificial Societies and Social Simulation 1(4) (1998),
http://www.soc.surrey.ac.uk/JASSS/1/4/1.html

A Model for HIV Spread in a South African Village

Shah Jamal Alam, Ruth Meyer, and Emma Norling

Centre for Policy Modelling,
Manchester Metropolitan University Business School
M13GH, Manchester, U.K.
{shah,ruth,emma}@cfpm.org

Abstract. This paper describes an agent-based simulation model of the spread of HIV/AIDS in the Sub-Saharan region. The model is part of our studying social complexity in the Sekhukhune district of the Limpopo province in South Africa. The model presents a coherent framework and identifies the essential factors agent-based modellers need to take into account when modelling HIV spread. The necessary empirical data are drawn from the villagers' accounts during our fieldtrip to the case study region and reports from the available epidemiological and demographic literature. The results presented here demonstrate how agent-based simulation can aid in a better understanding of this complex interplay of various factors responsible for the spread of the epidemic. Although the model is specific to the case study area, the general framework described in this paper can easily be extended and adapted for other regions.

Keywords: HIV/AIDS, evidence-driven modelling, sexual networks.

1 Introduction

This paper takes another step forward towards simulating the spread of HIV using agent-based modelling techniques. There have been several recent papers applying ABM to the spread of this pandemic ranging from those relying on simplistic assumptions to those driven by detailed epidemiological surveys and anecdotal evidence (c.f. [8, 12, 13, 19]). For modellers the greatest challenge is the availability of evidence, especially when the study is conducted in one of the most impoverished regions in the world: Sub-Saharan Africa. Another challenge for MABS modellers is determining the characteristics of dynamical networks as the simulation proceeds. If agents are heterogeneous and able to join or leave the modelled system during the simulation, the network characteristics may change completely over the course of a simulation run. Modelling dynamical populations of agents is necessary for exploring the transmission of HIV both horizontally and vertically. We adopt this methodology in this paper.

The human immunodeficiency virus (HIV) is passed on by the transmission of blood, semen, or breast-milk from an infected person. As with other sexually transmitted diseases (STDs) HIV is spread through intimate contacts between individuals. HIV can also be transmitted through infected syringes and via mother-to-child transfer. In the case of sexual intercourse, the spread is dependent upon how the involved

N. David and J.S. Sichmann (Eds.): MABS 2008, LNAI 5269, pp. 33–45, 2009.

individuals are sexually linked. In a monogamous relation, for example, the individuals are linked in a dyad and do not contribute to the spread of HIV (and other STDs) outside their relationship. The topology and dynamics of a sexual network depend upon the cultural context of the respective society. Tendency of clustering, mixing behaviour and the variation in the sexual contacts are important determinants of the network and have significant implications in the spread of the epidemic [7].

Few empirical studies are available that address the role of sexual networks in the spread of HIV/AIDS and other STDs in Sub-Saharan Africa. One such study is Helleringer and Kohler's report on individuals' preferences for sexual partners and the resulting sexual network on the Likoma Island in Malawi [3]. A major problem with such empirical data is the 'boundary-specification problem', which refers to the task of specifying inclusion rules for actors or relations in a network [6]. It is hard to validate if the studied social network is indeed representative of the examined phenomena. In other parts of the world, several studies have claimed the typical characteristic of a sexual network to be that of a scale-free network (c.f. [8, 9]). In the case of Sub-Saharan Africa, we lack the evidence whether the sexual network in rural areas follows a power-law distribution.

This paper describes the extended epidemiological component of an evidence-driven agent-based model of the impact of socioeconomic stressors, including HIV/AIDS, in a village in South Africa [1]. It is implemented in Java using the Repast simulation toolkit [25]. The epidemiological component has been enhanced with further evidence from several studies conducted in the region (c.f. [3, 18]) as well as the anecdotal accounts from villagers of Ga-Selala in the Limpopo province, South Africa, gathered during our own fieldwork.

In the next section, we introduce the model component specific to the spread of HIV/AIDS. Section 3 presents the simulation results, followed by a brief discussion on the related work in section 4. Section 5 presents the conclusions and outlook.

2 Model Specification

In this section, we describe the model parameters and processes related to the spread of HIV/AIDS in the simulated village society. As homosexuality is considered taboo in the case study area and empirical data are therefore lacking, the model so far considers only heterosexual relationships. We are aware that this may not reflect reality correctly and introduce a bias into our results.

2.1 Agents

Agents represent individuals characterized by their gender, age group, marital status, health status and expected lifetime. They live together in households, engage in sexual interaction, marry, work, migrate and maintain kinship and friendship links. Household composition is derived from detailed survey data of the Sekhukhune District made available to us by the RADAR[1] programme.

Stages from HIV to AIDS and Health Status. At the start of the simulation, a proportion of agents are initialized with HIV/AIDS based on gender and age-specific

---

[1] http://web.wits.ac.za/academic/health/PublicHealth/Radar/

prevalence data from the South African National HIV Survey[2]. As the simulation proceeds, agents transmit HIV via sexual intercourse (horizontal transmission) and from mother to child in the case of newborn agents (vertical transmission). There is also a possibility for migrant workers to contract HIV, which we discuss later on.

The stages of the disease from being infected with HIV to developing AIDS and finally dying have been modelled in various ways (c.f. [5, 16]). Our model follows the WHO clinical staging of HIV disease in adults and adolescents, which consists of four stages. In the first stage (primary HIV infection), it takes a few weeks until seroconversion, where an HIV infection can be detected. As the model runs on a monthly time scale, we have assumed this stage to pass in a month (time step). The second stage (clinically asymptomatic stage) lasts for a median of 8 years. In the model, this time-lag varies for each agent and is sampled using the Weibull distribution to get an overall central value of 8 years. The third and fourth stages describe the progression from symptomatic HIV infection to AIDS. We have combined these into one last stage where agents are identified with AIDS and progress to death in 10 to 13 months time. Figure 1 illustrates the stages in the model.

Adult agents continue their sexual activities until they are identified as having AIDS. In this last stage their health declines rapidly, forcing them to leave their current job, if any. Fertility of female agents declines from the clinical latency period until death. In future, we plan to incorporate the data on age-specific fertility available for rural South Africa[3].

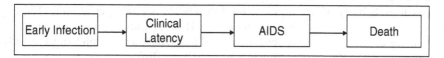

Fig. 1. Stages of progression to AIDS for agents in the model

Birth and Death of Agents. Child agents are born to couples or single mothers during the simulation. Pregnancy is only possible if the male partner is not away on migration. The male to female ratio of children is assumed 1:1. Agents are assigned an expected lifespan at the time of their creation. However, they may die much earlier due to infant mortality, AIDS, or lack of nutrition and can severely retard the population growth of the community [1].

2.2 Model Parameters for the Spread of HIV/AIDS

In this section, we discuss the parameters concerning the sexual mixing process, agents' preferences for sexual partners and the epidemiological aspects.

Distribution of Sexual Partners. Evidence shows that people in the case study area may have several sexual partners at the same time. We have used a lognormal distribution to model the maximal number of concurrent sexual partners, with different settings for male and female agents as reported by Helleringer and Kohler [3] in their

[2] http://www.avert.org/safricastats.htm
[3] e.g. http://www.hst.org.za/healthstats/5/data/geo

study of individual sexual activities and behaviour on the Likoma Island (Malawi). Unlike men, women in the region show a much lower tendency of having multiple sexual partners (see figure 2). Most female agents have at most one or two sexual partners at a time. The current model does not represent sex workers as individual agents.

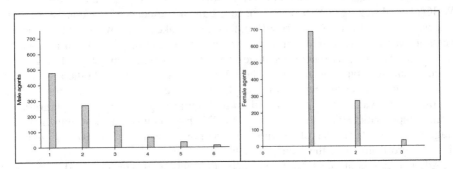

Fig. 2. Frequency distribution for maximum number of concurrent sexual partners for (left) male and (right) female agents

Transmission via Sexual Intercourse. The chance of the transmission of HIV per coital act has been reported variably in studies and clinical trials by epidemiologists. A typical probability of transmission used in models is 0.001 per coition [12, 13]. Nevertheless, the chance of the transmission of HIV infection varies with the stages of the disease. For example, Koopman et al. [5] have studied the implications of early transmission when this probability is highest. Using different probabilities for the stages of HIV can influence the rate of spread of the infection in the community. Table 1 gives the transmission probabilities for different stages used in the model. We have also tested a different set of values that are adapted from a study in by Wawer et al. in Uganda [17] (see section 3.2).

Since the model runs on a monthly timescale, an individual's chance of contracting the infection is calculated by multiplying the transmission probability with the number of coital acts/month (picked randomly from 7-13)[4]. A sexual partner of an infected migrant is only at risk whenever the migrated agent visits home for a 1-3 months break, every 4-6 months.

Table 1. Different transmission rates per coital from [5], used in the model

Stage	Duration	Transmission Probability
Early	3 months	0.2
Middle	104 months (8 years)	0.001
Late	10-13 months until death	0.076

Mother-To-Child Transfer and Child Mortality. Without treatment, there is typically 20-45% chance for a mother-to-child HIV transfer (MTCT). However, given the

---

[4] Wawer et al. [17] report an average of 10 coital acts/month.

availability of drugs and regular treatment, the chance can be reduced to as low as 2%. In the model we have assumed a 30% chance for MTCT without Antiretroviral (ARV) treatment [12]. Three different types of treatments based on the WHO guidelines are currently being considered in the model. They range from the minimal and least effective treatment of a single dose of nevirapine to the full treatment of mother and newborn with azidothymidine (AZT) for 28 weeks. The chance for MTCT varies considerably depending upon the available treatment; the accessibility of effective treatment therefore has significant policy implications.

As estimated by Newell et al. [10], 32.5% of the infants born with HIV infection die within a year and around 50% die within two years. Modelling infant mortality is imminent for studying HIV spread as this affects the sustainability of the community in successive generations. Table 2 gives the probability used in the model for HIV-related infant mortality up to 2 years of age. After that, a child follows the stages in the development of HIV as discussed before.

Table 2. HIV-related infant deaths in South Africa [10]

Year	Deaths due to HIV
≤1	325%
≤ 2	52.5%

2.3 A Simple Sexual Mixing Scheme

Modelling dynamic social networks requires a mechanism for sexual mixing of the agents. Sexual behaviours and preferences for choosing a sexual partner are strongly influenced by the social norms of the community [2, 14]. For our project, we have implemented several sexual mixing processes through which individuals form sexual relationships. The schemes assume a heterosexual society and monogamous marriage where male agents look for potential partners. A female agent may accept or reject the courtship offer. The schemes differ in terms of agents' choice criteria for a partner, courtship duration and the distribution of sexual partners. A comparison of the implemented schemes and the resulting sexual networks would go beyond the scope of this paper. For demonstrating the feasibility and usefulness of our approach, it is sufficient to focus on one of these schemes.

Attractiveness is a score assigned to agents (log-normally) and a male agent looks for a female whose attractiveness exceeds their own aspiration level (assigned log-normally as well). Agents without a single female partner have their aspiration levels decreased successively. For those satisfied with their current sexual partner(s), the aspiration level is increased incrementally.

In addition to the above rule, we have also implemented rules specific to rural South African communities. In this scheme, incest and inbreeding are prohibited. Agents search for sexual partners mostly among their friends and friends of friends. The friendship network evolves dynamically based on the rules outlined by Jin et al. [4] and informal savings clubs (stokvels) [1]. There is also a 5-10% chance for picking a female as potential partner randomly. This follows from the villagers' account during our fieldtrip to Sekhukhune District. Furthermore, young-adult agents look for partners of the same age group. The female agents' criteria change with age: Young female agents prefer males of similar age, while older female agents prefer unmarried suitors who have some employment.

The sequence of steps at each tick (simulation month) in the simple sexual mixing scheme adapted from [15] follows as:

```
1. For all adult male agents looking for sexual partners
      a. Update aspiration level
      b. If agent is not away or is visiting home
            Then find potential female partners
      c. Choose a female agent and send courtship offer
2. For all adult female agents
      a. Update aspiration level
      b. If agent has received an offer
            Then choose a suitor and reject the rest
3. For every unmarried couple, count courtship duration
```

There is a courtship time in the model for each agent [15, 19]. From the anecdotal evidence we know that it takes 1-2 years before a couple decides to marry. Since a male has to pay *lobola* (bride price) to the female's household for the right to marry her, the marriage of the couple can be further delayed [1].

Not every sexual relationship ends in marriage, though. So far, we have modelled the break-up of a partnership randomly, using different probabilities. We are planning to incorporate more adequate break-up rules from a study in the region [3]. Finally, we are also exploring the effects of the assumption that married female agents do not have sexual partners outside of their marriage.

3 Preliminary Simulation Results

In the simulation experiments for this paper, we have set the initial size of the modelled community (a village in the Sekhukhune district, Limpopo province, South Africa) to be 111 households (~950 individual agents). At each time step, we introduce new cases of HIV for migrant agents using a Gamma distribution [1]. This accounts for some migrant agents having sexual interaction with sex workers.

3.1 Searching Potential Female Sexual Partners

From the anecdotal evidence, we know that sexual partners usually are already acquainted by some means or other but to some extent may have met by chance. To investigate the effect of the proportion of random encounters on the spread of HIV, we conducted experiments for three different values with the default parameter values and sexual mixing mechanisms describe above. The upper age limit for sexual activity was set to 55 for male and 45 for female agents. At each time step, male agents choose female partners from their circle of friends and acquaintances, or with a 5, 15, or 25% chance choose from all adult females. Figure 3 shows the average proportion of agents infected with HIV of 5 runs each for 900 steps (~75 years).

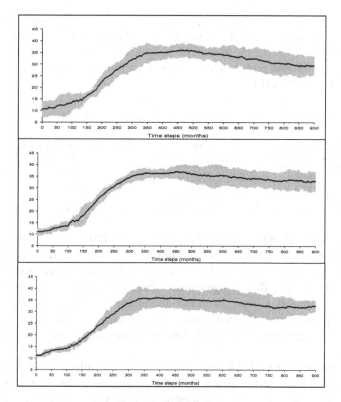

Fig. 3. Average percentage of HIV infected agents with (above) 5%; (middle) 15% and (bottom) 25% chance for a random search, with 2nd standard deviation

As can be expected, the percentage of HIV infected agents is lowest in the first case (i.e. with 5%). This is because agents mostly stay within their social groups. With more opportunities for encountering a female agent randomly, the spread of infection increases slightly. On average, the development over time is quite similar in all three cases. However, the pattern of variability in the simulations runs differs especially for the first half of the runs. This indicates that with a higher probability for finding a sexual partner, almost all the female agents of marriageable age were approached by the male agents. The variability increases in the later half, as the population decreases due to deaths and a decrease in the number of births. This results in fewer female agents available for new sexual partnerships.

3.2 Role of Different HIV Transmission Rates and Exogenous Incidence

Different probabilities for HIV transmission per coital act for the different stages of HIV have been reported. In the previous section, we applied the probabilities reported by [5] with quite high rates of transmission in the early and late stages. To explore how the transmission rates affect the observed HIV prevalence, we combined three different sets of rates with the introduction of new cases of HIV infection amongst migrants ('exogenous incidence'). The investigated cases are: (I) transmission rates as reported in [5] with exogenous incidence and (II) without, (III) transmission rates as reported in [17] with exogenous incidence and (IV) without, and (V) a constant transmission rate of 0.0012 [5, 9] without exogenous incidence. Table 3 shows how the average transmission probabilities per coital act from [17] have been adapted into the model for cases III and IV.

Table 3. Different transmission rates per coital act adapted from [17]

Stage	Duration	Transmission Probability
Early	3 months	0.0082
Middle	Until 15 months	0.0015
Middle	90 months	0.0007
Late	14 months until death	0.0028

As before, selected agents are initialized with HIV/AIDS based on their gender and age at the start of the simulation. In contrast to the simulation runs reported in the previous section, we now also randomly assign each infected agent with one of the different stages of HIV. This change causes a large proportion of agents to die within the first 10-15 simulation years. Note that agents are not assigned with extra-marital sexual partners at the start. Consequently, transmission via sexual interaction drops in all cases in the beginning of the simulation and remains on a lower level than before (see section 3.1).

The results shown in Figure 4 clearly demonstrate the influence of both exogenous incidence and different probabilities used for HIV transmission on the overall percentage of HIV infected agents. Introducing new infections from 'outside' into the simulated population keeps the epidemic going (see case I and III) whereas without such exogenous incidence it eventually runs its course, with prevalence declining to zero (case IV and V). The higher transmission rates in case II causes the decline to take longer as more agents have been reached by the spread of HIV.

The transmission rates play a large role in determining the extent of the spread. Using the rather high probabilities reported by Koopman et al. [5] (see Table 1) results in a significantly higher average prevalence, even without exogenous incidence. For the set of probabilities from a more recent study in Uganda [17] (see Table 3), the average prevalence is low compared to the first two cases. The effect of death of initially infected agents in the early phase is much more prominent with low transmission probabilities. Case V with its constant transmission rate and lack of exogenous incidence can be regarded as a lower limit for the simulated system.

The five cases highlight an important cause of HIV spread in the Sub-Saharan region, which is the role of commercial sex workers and migrant agents. Currently, this is introduced as a macro component modelling the exogenous incidence. Also, one needs to further explore the effects of different probabilities at different stages as they have important consequences with respect to AIDS prevention measures [5]. Last but not least, sexual transmission holds the key for HIV transmission for all settings and is influenced by our choice of the sexual mixing scheme. The effects of the latter will be explored as next steps.

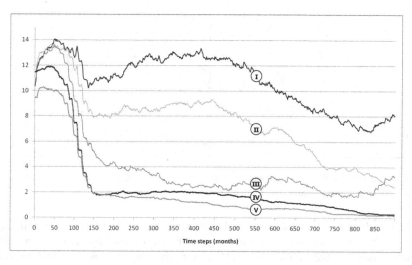

Fig. 4. Average percentage of HIV infected agents for the five cases

3.3 Characteristics of the Sexual Network

The sexual network in our model is dynamic in that not only links are formed and dropped again based on the model parameters and the mixing scheme discussed before but also nodes (agents) enter and leave the network over time. Since only heterosexual relations are considered in the model, we get a 2-mode network. 2-mode networks have received considerable attention recently in the social network analysis (SNA) community and several alternative counterparts of the network analysis measures (e.g. clustering) have been proposed. Moreover, analysing networks with changing populations is still challenging for SNA researchers.

A spanning tree structure is a typical signature of sexual activities in a heterosexual community [c.f. 22]. The empirical research by [22] shows the community of college students to be linked in a single spanning tree. A chainlike spanning tree is characterized by few cycles, low redundancy and sparse density. Note that a spanning tree is not possible in case of random mixings and positive preferences for partners [22]. Such characteristics evolve only in the presence of both positive and negative preference rules when individuals may choose to accept or reject a relationship.

With the current model setup, we get a spanning forest instead of a spanning tree. A spanning forest has several definitions; in our context, it means a forest of disconnected spanning trees. Figure 5 shows the disconnected spanning trees of the sexual network for two different settings, the default distribution for the maximum number of concurrent partners (see Figure 3) and a distribution with increased number maximum sexual partners allowed.

In Figure 5 (left) we see only a few spanning-tree-like structures compared to Figure 5 (right). Dyads have been omitted for clarity in the visualization. These dyads are mainly due to marriages where a female agent is limited to marital sexual activity and her husband is unable or unwilling to find additional sexual partners. Increasing the number of allowable concurrent partners decreases the dyad frequency and increases the possibility of higher sub-graph structures.

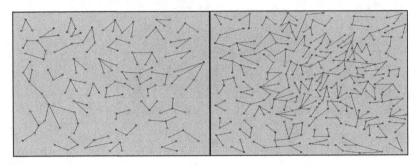

Fig. 5. Snapshot of sexual network taken at the 350th tick for (left) the default maximal numbers of concurrent partners and (right) increased numbers

The simple sexual scheme discussed in this paper does not lead to the joining of many small-sized spanning trees into long chains as reported by [22]. Nevertheless, the individuals' choice for partners is reflected in the snapshot taken from the simulation. Obviously, as the network changes at every time step, looking at just a single snapshot may not be enough to understand the underlying processes. A detailed study of these simulated heterosexual networks and the role of different mixing schemes in their evolution will be presented separately.

4 Some Related Work

Using agent-based simulation to model the spread of HIV is still a rather novel approach. To our knowledge only a few other agent-based models exist today, which limits the opportunity to put our own work in context. Notice that many microsimulation and compartmental models have been developed in the past concerning the spread of HIV/AIDS and its impact. In this section, we restrict our discussion to some of the agent-based models for the spread of HIV/AIDS. Rhee [13] presents an agent-based model of HIV spread and policy interventions based on a community in Papa New Guinea. Agents maintain sexual and friendship links through which the idea of condom usage is spread. The model uses a fixed transmission probability and takes care of the different stages from HIV to AIDS. The population is kept fixed and HIV is spread through a stochastic sexual mixing scheme [21]. Rhee reports a prevalence of ~30% on a scale-free network in ~20 years. This is similar to the results reported in this paper, although in our case the epidemic plateau undergoes changes due to births and deaths of the agents.

Sumodhee et al. [16] have looked into the role of social behaviour on the spread of HIV in a Taiwanese homosexual community. The purpose of the model was to assess factors like 'number of partners per agent', 'condom usage' and 'degree of faithfulness towards long-term partners'. The model reproduces an artificial social network with scale-free characteristics and focuses on individual behaviour. They assume different courtship durations within and across groups of similar sexual activities and the resulting sexual network changes as new links are formed and existing ones are broken. In contrast to [16], in our model, we assume a heterosexual community. Moreover, we have not yet incorporated the use of contraceptives into the model. Courtship duration in our case differ across individuals and depend upon the number of current partners.

Heuveline et al. [19] outlined an agent-based model, building upon a microsimulation model, aimed at understanding the demographic impact of HIV/AIDS. The proposed local processes related to marriage patterns, migration, horizontal and vertical transmission of HIV and social interaction among the agents. Their model thus addresses issued related in our overall model [1] and in this paper. The authors planned to incorporate demographic data into their model as we have used in our case. However, the social aspects covered in our model reflect stakeholders' perceptions from a real case study from the CAVES Project (see [1, 23]). Unfortunately, any further development by the Heuveline et al. is not been reported.

The model reported in a recent paper by Tawfik and Faraj [24] simulated a rural society in the Limpopo province, South Africa. The model incorporates marriages among agents resulting in the creation of new households during the simulation, which become part of the social network. While their paper is similar to our previous work [1], they do not take into account the economic impact of HIV/AIDS. Instead, their work is directed towards the use of social network ties in propagating awareness and applying various intervention strategies against HIV/AIDS and tuberculosis. However, they do not the different stages of the HIV infection and the change in transmissibility as HIV-positive agents progress from one stage to another.

5 Conclusions and Outlook

Understanding the impact of HIV/AIDS is a multidimensional problem, and a model must be constrained by evidence wherever possible. Traditional models of HIV spread based on Monte Carlo simulations and system dynamics techniques (e.g. [12]) have been developed for almost two decades. However, such techniques are typically infeasible in tackling the interplay of multiple processes especially when dealing with complex social systems.

To our knowledge, agent-based simulation models still lag behind in successfully capturing the recent advances in understanding HIV transmission. Most models address this problem with specifically assuming *a priori* network configurations (e.g. a scale-free or a small-world network [18]). However, having a dynamical sexual network where the links, i.e. the individuals' sexual partnerships, change during simulation in not enough. Firstly, the network characteristics do not necessarily remain stable in a dynamical population [20] and secondly, simulating vertical transmission of HIV is not possible.

As next steps, we would like to explore the model's parameters and its robustness under different initial settings. We are also investigating the role of different sexual mixing processes about the evolved network characteristics and HIV transmission.

Incorporating further evidence can lead towards a better understanding of the long-term impact of HIV/AIDS. There is a tendency for polygyny in the region, which will be implemented next. Furthermore, we need to incorporate rules to restrict agents to at most one sexual partner towards the later stages of the disease [11]. Lastly, HIV infectivity can increase significantly in the presence of other STDs and diseases such as genital ulcers, and we would like to address this as future work.

Acknowledgements. This work was supported under the EU FP6 Project CAVES. We are thankful to James Koopman and the two anonymous referees for their feedback. We would also like to thank Scott Moss, Bruce Edmonds, the RADAR programme, and our case study partners from the Stockholm Environment Institute (Oxford), in particular, Gina Ziervogel, Anna Taylor and Sukaina Bharwani.

References

1. Alam, S.J., Meyer, R., Ziervogel, G., Moss, S.: The Impact of HIV/AIDS in the Context of Socioeconomic Stressors: An Evidence-driven Approach. Journal of Artificial Societies and Social Simulation 10(4), 7 (2007),
   http://jasss.soc.surrey.ac.uk/10/4/7.html
2. Billari, F., Prskawetz, A., Fürnkranz, J.: The evolution of social norms: age norms on marriage. Max Planck Institute for Demographic Research, Rostock, Germany (2005)
3. Helleringer, S., Kohler, H.-P.: Sexual Network Structure and the Spread of HIV in Africa: Evidence from Likoma Island, Malawi. AIDS 21(17), 2323–2332 (2007)
4. Jin, E.M., Girvan, M., Newman, M.E.J.: Structure of growing social networks. Phys. Rev. E. 64.4 (2001)
5. Koopman, J.S., et al.: The role of early HIV infection in the spread of HIV through populations. J. Acq Imm. Def. Syn. Hum. Retroviral 14, 249–258 (1997)
6. Kossinets, G.: Effects of missing data in social networks. Social Networks 28 (2006)
7. Liljeros, F., Edling, C.R., Amaral, L.A.N.: Sexual networks: implications for the transmission of sexually transmitted infections. Microsbes Infect., 189–196 (2003)
8. Liljeros, F., et al.: The Web of Human Sexual Contacts. Nature 411, 908–909 (2001)
9. Merli, M.G., et al.: Modelling the spread of HIV/AIDS in China: The role of sexual transmission. Population Studies 60(1), 1–22 (2006)
10. Newell, M., et al.: Mortality of infected and uninfected infants born to HIV-infected mothers in Africa: a pooled analysis. The Lancet. 364, 1236–1243 (2004)
11. Nordvik, M.K., Liljeros, F.: Number of Sexual Encounters Involving Intercourse and the Transmission of Sexually Transmitted Infections. Sex. Trans Dis. 33(6), 342–349 (2006)
12. Rauner, M.S., et al.: Use of Discrete-Event Simulation to Evaluate Strategies for the Prevention of Mother-to-Child Transmission of HIV in Developing Countries. J. Operational Research Society 56, 222–233 (2005)
13. Rhee, A.: An Agent-based Approach to HIV/AIDS Modelling: A Case Study of Papua New Guinea. Master of Science Thesis. Massachusetts Institute of Technology (2006)
14. Schmitt, D.P.: Sociosexuality from Argentina to Zimbabwe: A 48-nation study of sex, culture, and strategies of human mating. Behavioral and Brain Sciences 28(2), 247–311 (2005)
15. Simão, J., Todd, P.M.: Emergent Patterns of Mate Choice in Human Populations. Artificial Life 9(4), 403–417 (2003)
16. Sumodhee, C., et al.: Impact of Social Behaviors on HIV Epidemic: A Computer Simulation View. In: Proc. Intl. Conference on Computational Intelligence for Modelling, Control and Automation, pp. 550–556. IEEE Press, Los Alamitos (2005)
17. Wawer, M.J., et al.: Rates of HIV-1 Transmission per Coital Act, by Stage of HIV-1 Infection, in Rakai, Uganda. Journal of Infectious Diseases 191, 1403–1409 (2005)
18. Watts, D.J., Strogatz, S.: Collective Dynamics of 'small world' networks. Nature 393, 440–442 (1998)
19. Heuveline, P., Sallach, D., Howe, T.: The Structure of an Epidemic: Modelling AIDS Transmission in Southern Africa. In: Papers from Symposium on Agent-based Computational Modelling, Vienna, Austria (2003)

20. Alam, S.J., Edmonds, B., Meyer, R.: Identifying Structural Changes in Networks Generated from Agent-based Social Simulation Models. In: Proc. 10th Pacific RIM International Workshop on Multi-agent Systems (2007)
21. Kretzschmar, M., Morris, M.: Measures of concurrency in networks and the spread of infectious disease. Math. Biosci. 133(2), 165–195 (1996)
22. Bearman, P.S., Moody, J., Stovel, K.: Chains of Affection: The Structure of Adolescent Romantic and Sexual Networks. American Journal of Sociology 110(1), 44–91 (2004)
23. Ziervogel, G., Taylor, A.: Feeling Stressed: Integrating Climate Adaptation with Other Priorities in South Africa. Environment 50(2), 32–41 (2008)
24. Tawfik, A.Y., Farag, R.R.: Modeling the Spread of Preventable Diseases: Social Culture and Epidemiology. In: Proc. IFIP International Federation for Information Processing; Artificial Intelligence and Practice II, vol. 276, pp. 277–286. Springer, Boston (2008)
25. North, M.J., Collier, N.T., Vos, J.R.: Experiences Creating Three Implementations of the Repast Agent Modeling Toolkit. ACM Transactions on Modeling and Computer Simulation 16(1) (2006)

Understanding Collective Cognitive Convergence

H.V. Parunak, T.C. Belding, R. Hilscher, and S. Brueckner

NewVectors division of TTGSI
3520 Green Court, Suite 250
Ann Arbor, MI 48105 USA
+1 734 302 4684
{van.parunak,ted.belding,sven.brueckner,
rainer.hilscher}@newvectors.net

Abstract. When a set of people interact frequently with one another, they often grow to think more and more along the same lines, a phenomenon we call "collective cognitive convergence" (C^3). We discuss instances of C^3 and why it is advantageous or disadvantageous; review previous work in sociology, computational social science, and evolutionary biology that sheds light on C^3; define a computational model for the convergence process and quantitative metrics that can be used to study it; report on experiments with this model and metric; and suggest how the insights from this model can inspire techniques for managing C^3.

1 Introduction

When the same people interact frequently with one another, the dynamics of the *collective* can lead to a *convergence* in *cognitive* orientation, thus "collective cognitive convergence" (C^3). C^3 is seen in many different contexts, including research subdisciplines, political and religious associations, and persistent adversarial configurations such as the cold war. Tools that support collaboration, such as blogging, wikis, and tagging, make it easier for people to find and interact with others who share their views, and thus may accelerate C^3. This efficiency is sometimes desirable, since it enables a group to reach consensus more quickly. For instance, in the academy, it enables coordinated research efforts that may accelerate the growth of knowledge.

But *convergence* can lead to *collapse*. It reduces the diversity of concepts to which the group is exposed and thus leaves the group vulnerable to unexpected changes. For example, in academia, specialized tracks at conferences sometimes become unintelligible to those who are not specialists in the subject of a particular track, and papers that do not fit neatly into one or another subdiscipline face difficulty being accepted. The subdiscipline is increasingly sustained more by its own interests than by the contributions it can make to the broader research community or to society at large.[1]

Groups that have undergone cognitive collapse will only produce output conforming to their converged set of ideas, and will be unable to conceive or explore new

---

[1] This paper was motivated by frustration in the industry track at AAMAS07 that some subdisciplines of agent research were becoming so ingrown, focusing only on problems defined by other members of the subdiscipline, that it was difficult to apply them to real problems.

N. David and J.S. Sichmann (Eds.): MABS 2008, LNAI 5269, pp. 46–59, 2009.

ideas. In the worst case, self-reinforcing collapse may lead a group to focus on a cognitive construct meaningful only to the group's members. Highly specialized academic disciplines can become increasingly irrelevant to people outside of their own circle.

We became interested in this phenomenon by observing increasing balkanization in multi-agent research. Since we work in multi-agent simulation, it occurred to us that some light might be shed on the phenomenon with a multi-agent model. Many groups do not collapse, and a simulation can also help us understand how they maintain their diversity. This paper presents some preliminary results.

Section 2 discusses previous work related to our effort. Section 3 describes our model, and a metric that we use to quantify C^3. Section 4 outlines a series of experiments that exhibit the phenomenon and explore possible techniques for managing it. Section 5 suggests directions for further research, and Section 6 concludes.

2 Previous Work

Our research on C^3 builds on and extends previous work in sociology (both empirical and computational) and evolutionary biology.

2.1 Sociological Antecedents

There is abundant **empirical** evidence that groups of people who interact regularly with one another tend to exhibit C^3. Sunstein [27] draws attention to one version of this phenomenon, "group polarization": a group with a slight tendency toward one position will become more extreme through interaction. This dynamic suggests that confidence in group deliberation as a way of reaching a moderating position may be misplaced. He summarizes many earlier studies, and attributes the phenomenon to two main drivers: social pressure to conform, and the limited knowledge in a delimited group. Our model captures the second of these drivers, but not the first. Sunstein suggests some ways of ameliorating the problem that we explore with our model.

One recent review of **computational** studies of consensus formation [14] traces relevant studies back more than 50 years [11], including both analysis and simulation. They differ in the belief model and the topology, arity, and preference of agent interactions. Rather than attempting an exhaustive review, we situate our work in these dimensions.

Belief.—An agent's belief can be either a single variable or a vector, with real, binary, or nominal values. Vector models usually represent a collection of beliefs, but in one study [3] the different entries in the vector represent the value of the same belief that underlies different behaviors, to explore of internal consistency.

Topology.—Some models constrain interactions by agent location in an incomplete graph, usually a lattice (though one study [18] considers scale-free networks). In others any agents can interact (often called the "random choice" model).

Arity.—Agents may interact only two at a time, or as larger groups.

Preference.—The likelihood of agent interaction may be modulated by their similarity.

Table 1. Representative Studies in Consensus Formation

Study	Belief	Topology	Arity	Preference?
Krause [17]	Real variable	Random	Many	Yes
Sznajd-Weron [28]	Binary variable	Lattice	Two	No
Deffuant [7]	Real variable	Random	Two	Yes
	Binary vector	Random	Two	Yes
Axelrod [2]	Nominal² vector	Lattice	Two	Yes
Bednar [3]	Nominal vector³	Random	Many	No
This paper	Binary vector	Random	Many	Yes

Table 1 characterizes several previous papers in this area in terms of these dimensions. Our work represents a unique combination of these characteristics. In particular,

- We consider a vector of m beliefs, rather than a single belief. This model allows us to look at how an individual may participate in different interest groups based on different interests, but also makes describing the dynamics much more difficult than with a single real-valued variable. In the latter case, individuals move along a linear continuum, and measures such as the mean and variance of their position are suitable metrics of the system's state. In our case, they live on the Boolean lattice $\{0,1\}^m$ of interests, and our measures must reflect the structure of this lattice.
- We allow many individuals to interact at the same time. This convention captures the dynamics of group interaction more accurately than does pairwise interaction, but also means that our agents interact with a probability distribution over the belief vector rather than a single selection from such a distribution.
- We allow our agents to modulate the likelihood of interaction based on how similar they are to their interaction partners. This kind of interest-based selection is critical to the dynamics of interest to us, but makes the system much more complex.

One consequence of selecting the more complicated options along these dimensions is that analytic results, accessible with simpler models, become elusive. Almost all analytical results in this discipline are achieved by modeling the belief of agent i as a single real number x_i and studying the evolution of the vector x over time as a function of the row-stochastic matrix A whose elements a_{ij} indicate the weight assigned by agent i to agent j's belief, $x(t+1) = Ax(t)$. This model captures interaction arity greater than two, but not vector beliefs or agent preferences. Conditions for convergence under preferences have been obtained [17], but only for six or fewer agents [14]. Bednar et al. [3] have derived convergence times for a form of vector belief, but only for binary interactions and with no preferences. Even for binary interactions, the combination of vector-based beliefs and preferences has resisted analytical treatment (in studies of an isomorphic system, bisexual preferential mating [15, 24]).

In a forthcoming paper [21], we have derived formal convergence results for a very simple case of our model (all agents interacting as a single group), but even this model is unable to capture the full richness of behavior that we observe. Given this

---

² [10] finds faster convergence when some elements in the vector function as interval variables.
³ All entries reflect the same belief in different behavioral settings, and pressure toward internal consistency is part of the model dynamics.

research context, in this paper we focus our attention on simulation results, to develop intuitions that may reward future analytical exploration.

2.2 Biological Antecedents

The subgroups that form and cease to interact when convergence turns to collapse resemble biological species, which do not interbreed. So we look for insight to research in biological speciation (see [5, 12] for reviews). Compare Dawkins' notion of the meme in cultural evolution [6]. The most commonly proposed speciation mechanisms are allopatric, sympatric, and parapatric.

- In allopatric speciation, genetic barriers gradually evolve between two or more geographically isolated species (for instance, organisms living on separate islands). One configuration of our model can be interpreted as allopatric speciation.
- Parapatric speciation has no discrete barrier between populations; individuals are distributed along a geographic continuum and are separated by distance. Finally, in sympatric speciation a single population with no physical or geographic gene flow barriers divides into separate species.
- Sympatric speciation requires two interacting forces: 1) a force that drives sympatric speciation (e.g. resource competition or sexual selection) and 2) assortative mating that generates phenotypic variability and maintains evolving phenotypic clusters that eventually become species. Assortative mating refers to a mating system where different individuals express preferences for different phenotypes (e.g. some female birds prefer males with red feathers and other females prefer males with blue feathers). Some configurations of our model correspond to sympatric speciation.

Sexual selection [1, 9] refers to the differential mating success of individuals in a population, and can be based on either an asymmetric mating system (males compete and females choose) or a symmetric mating system (mutual mate choice where both sexes compete and choose). One sexual selection mechanism, Fisher's runaway process, leads to extravagant traits in males that are detrimental to their survival.

While the relative importance and frequency of these speciation mechanisms in nature are still heavily debated, the mathematical prerequisites for each mechanism have been extensively studied [5, 12, 16]. This work could be adapted to predict when and how C^3 will develop, and how it can be managed.

Our C^3 model can be considered an instance of a runaway sexual selection speciation model with mutual mate choice. We assume a homogenous environment, no physical barriers for the exchange of ideas and a symmetric "mating system" where individuals express their "mating preferences" (i.e. their preference for an atomic interest; see Section 3 below) mutually. In our model, a preference for extreme traits is modeled as the probability of adopting an interest based on the prevalence of this interest in a given neighborhood. A successful runaway process in our model can be viewed as the development of academic specializations with little practical relevance.

There has also been much theoretical work done to study the amount of gene flow or migration that is necessary to prevent isolated populations of organisms from diverging or losing diversity due to genetic drift, or sampling error [13]. Sewall Wright argued in his Shifting Balance Theory that a subdivided population with intermittent migration could exhibit more rapid evolutionary change than a single cohesive breeding population [25]. The mathematical frameworks for studying migration could be applied to modeling the exchange of ideas or individuals between groups in C^3, and the amount of exchange that is necessary to prevent intellectual isolation.

3   A Model and Metrics

We have constructed a simple multi-agent model of C^3 to study this phenomenon. Each participant's interests are a binary vector. Each position in the vector corresponds to an atomic interest. A '1' at a position means that the participant is interested in that topic, while a '0' indicates a lack of interest. Similarity between two vectors is measured by the normalized Hamming similarity, which is the number of positions at which the two vectors agree, divided by the overall length of the vectors. At each step, each participant

- identifies a neighborhood of other participants based on some criteria (here, usually similarity between their interest vectors greater than a similarity threshold θ, but alternatively geographical proximity, or proximity in a social network),
- learns from this neighborhood (by changing an interest j currently at 0 to 1 with probability $p_{interest}$ = proportion of neighbors having interest j set to 1), and
- forgets (by turning off an interest j currently at 1 to 0 with probability $1 - p_{interest}$).

One boundary condition requires attention. If an agent has no neighbors, what should $p_{interest}$ be? We take the view that interests are fundamentally social constructs, persisting only when maintained. Thus an isolated agent will eventually lose interest in everything, and in our model, a null community leads to $p_{interest} = 0$ for all interests. Alternative assumptions are certainly possible, and would lead to a different model.

We need quantitative measures of agent convergence to study C^3 systematically. Because we are working with vectors of beliefs rather than a single belief, simple summaries such as the mean and variance of a scalar (commonly used in studies of the system $x(t+1) = Ax(t)$) are not available to us. We have examined a number of possible measures over sets of belief vectors.

- Bednar et al. [3] deal with belief vectors, but where the elements reflect (possibly inconsistent) manifestations of the same belief under different circumstances. The attractor of interest for a single agent is thus a vector all of whose entries are the same, and for a community, concurrence across these single values.
- Some speciation models that model genomes as Boolean vectors [15] use the number of shared 1's in the vectors x_j, x_k of two agents a_j, a_k to measure their similarity, and average this quantity over a population to estimate the closeness of the population. This measure has the property that two vectors of all zero's are maximally separate; we prefer a measure (such as the Hamming distance) that recognizes identical vectors as maximally similar, whether the agreement is in 1's or 0's. Simply averaging the similarity loses the important distinction between highly clustered agents and uniformly distributed agents.
- Sophisticated statistical techniques exist for estimating the "true" social beliefs of a population empirically, based on their (noisy) responses to questionnaires [26]. In our case, the vectors are accurate representations of the agents' beliefs.

- One interesting class of measure that we have not pursued, and that might usefully supplement our measures, is the persistence of agent associations over time [20].

In the original version of this research [22], we derived a convergence measure from the distances at which agents cluster in a hierarchical clustering. This measure is costly to compute and does not lend itself to analytical treatment. A much more satisfactory measure is the mutual information between topics and agents. The mutual information between two features (a, b) over a set of data is

$$MI = \sum_{a,b} p(a,b) \log\left(\frac{p(a,b)}{p(a)p(b)} \right)$$

In our case, b indexes over topics and a indexes over agents. $p(a,b)$ is the probability that the ath agent is interested in the bth topic, while $p(b)$ is the sum of this probability over all agents, and $p(a)$ is the sum for agent a over all topics. Consider two cases.

First, if all agents have the same interests, $p(a)$ is independent of $p(b)$. Then $p(a,b) = p(a)p(b)$, and the logarithm (and thus MI) vanishes.

Second, if each agent has a distinctive set of interests, $p(a,b)$ will differ from $p(a)p(b)$. Sometimes it will be less, sometimes more. When it is less, the logarithm will be negative, but will be weighted by a relatively small value of $p(a,b)$. When it is more, the logarithm will be positive, and will be weighted by the larger $p(a,b)$. The resulting MI will be greater than zero, with an upper bound equal to the lesser of $log(m)$ and $log(n)$.

Thus MI is an easily computed measure of the degree of diversity in an agent population. 0 indicates that all agents have identical interests, while a higher value indicates divergence.

Fig. 1 shows the behavior of the MI over a sample run with 20 agents and interest vectors of length 10, where the probability of learning and forgetting is equal, and where agents are considered to be in the same group if the similarity between their interest vectors is greater than $\theta=0.5$. All agents collapse to the same interest vector at about generation 600. A generation consists of selecting one agent, choosing its neighbors, choosing

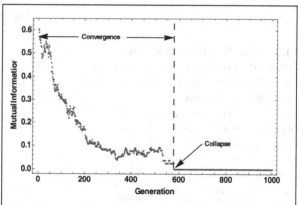

Fig. 1. Evolution of 20 agents, 10 interests, neighborhoods defined by similarity > 0.5

with equal probability whether it shall attempt to learn or forget, selecting a bit in its interest string at random, then if it is learning and the bit is 0, flipping the bit with probability $p_{learn} * p_{interest}$, or if it is forgetting and the bit is on, flipping the bit with probability $p_{forget} * (1 \ p_{interest}.)$.

4   Some Experiments

With this model and metric, we can explore C^3 under a variety of circumstances. As we might expect, forming groups based on similar interests leads to rapid cognitive convergence. But other sorts of neighborhoods also lead to convergence.

4.1   Things that Don't Work

We have explored several different kinds of neighborhood formation policies that also lead to convergence.

4.1.1   A Universal Group

Perhaps highly tolerant agents, who consider all agents their neighbors, might avoid convergence. Fig. 2 shows the evolution when two agents consider one another neighbors if their similarity is greater than 0 (that is, they have at least one bit position in common). This configuration might model a conference with only plenary sessions. The population still collapses.

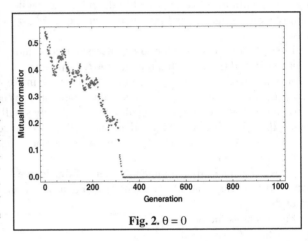

Fig. 2. θ = 0

4.1.2   Fixed Neighborhood Size

Perhaps the problem is that as agents converge, their neighborhoods in-crease in size. Fig. 3 shows the effect of defining an agent's neighborhood at each turn as the group of four other agents that are closest to it. This configuration models a conference with separate tracks, organized on the basis of the common interests of their members. It corresponds to the biological model of sympatric speciation. The assortative component is provided by the preference for partners with similar interests, while the limit on group size provides pressure toward diversity. Though agents base their

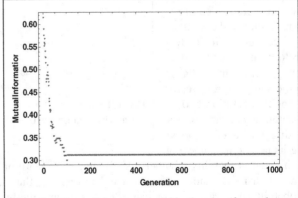

Fig. 3. Fixed-size agent neighborhoods (four closest agents)

adaptation at each turn on only 20% of the other agents, the *MI* still goes to zero, as agents form subgroups within which interests collapse.

4.1.3 Random Neighborhoods
Even more radically, let an agent's neighbors at each step be four randomly chosen agents (like a conference where papers are assigned to tracks, not by topic, but randomly). In spite of the resulting mixing, the population again collapses.

The time to convergence is highly variable, not only among these examples but within a single configuration.[4] The one constant across all runs is that the system does converge, in fewer than 500 generations (often far fewer).

4.2 Introducing Variation

The collapse of agent interests is abetted by the lack of any mechanism for introducing variation. Once the population loses the variation among agents, it cannot regain it. We have explored three mechanisms for adding variation to the population: random mutation, curmudgeons, and fixed interacting subpopulations.

4.2.1 Random Mutation
The simplest approach is mutation. At each generation, with some small probability p_{mutate}, after learning or forgetting, the active agent selects a bit at random and flips it. Fig. 4 shows an extended run with parameters the same as Fig. 1 (neighborhoods defined by $\theta = 0.5$), but with $p_{mutate} = 0.06$. Mutation is certainly able to reintroduce variation, but the level is critical. If mutation is too low (say,

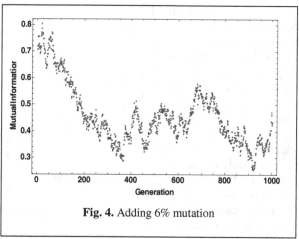

Fig. 4. Adding 6% mutation

1%), it is unable to keep up with the pressure to convergence, while if it is too high (10%), the community does not exhibit any convergence at all (and in effect ceases to be a community). The nature of its contribution follows a clear pattern. When it is in the critical range, the system occasionally collapses, but then discovers new ideas that reinvigorate it.

Mutation as an abstract mechanism corresponds to several possible effects in real-world group dynamics.

---

[4] This variation also shows the limits of applying a simple analytical derivation of convergence time such as that of Bednar et al. [3] to our more complex system. For the simplest case of a universal group, a forthcoming study [21] shows that convergence is proportional to $e^{-t/mn}$, where m is the number of topics and n the number of agents, but this analysis does not take into account more complex groupings.

- An agent might exhibit spontaneous curiosity about some topic that it has not previously thought important.

- An agent's attention might be drawn to a topic because of exogenous events such as news reports. In effect, the agent is in multiple groups concurrently, the community whose members are modeled by the C^3 agents, and a broader community (e.g., subscribers to the *New York Times*). We model this interplay of information from multiple communities explicitly later in this section.

- Yves Demazeau[5] has suggested the dynamic of a changing set of topics over time. Mutation approximates this dynamic, in the following sense. Mutation replaces a single agent's position on a topic by a randomly chosen 0 or 1. If the population adds a new topic, concurrently deleting an old one so that the total number of topics remains the same, the effect is the same as if each agent were to mutate the same position in its vector, so that agents have random values in that position. Thus topic addition can be viewed as mutation across the whole population rather than just a single agent.

4.2.2 Curmudgeons

A curmudgeon is a non-conformist, someone who regularly questions the group's norms and assumptions. Sunstein [27] observes that "group members with extreme positions generally change little as a result of discussion," and serve to restrain the polarization of the group as a whole.

Recall that ordinarily agents learn by flipping a 0 bit to 1 with probability $p_{interest}$ (the proportion of neighbors that have the bit on), and forget by flipping a 1 bit with probability equal to $1 - p_{interest}$. To model curmudgeons, when an agent decides to learn or forget, with prob-

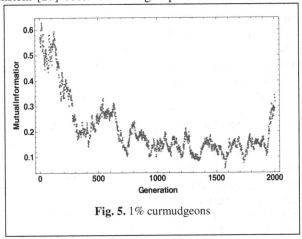

Fig. 5. 1% curmudgeons

ability p_{cur}, it reverses these probabilities. That is, its probability of forgetting when it is curmudgeonly is $p_{interest}$ (instead of $1 - p_{interest}$ in the non-curmudgeonly state), and its probability of learning is $1 - p_{interest}$.

Fig. 5 shows the effect when 1% of the agent decisions are reversed, again with the baseline configuration of Fig. 3. The system clearly converges, but does not collapse. Furthermore, p_{cur} can achieve this balancing effect over a much wider range than p_{mutate}. As much as researchers may resent reviewers and discussants who "just don't get it," curmudgeons are an effective and robust way of keeping a community from collapsing, as long as they are not excluded from the community.

---

[5] In very helpful discussion at the MABS workshop where this paper was presented.

4.2.3 Fixed Interacting Subpopulations

The third source of variation is even more robust, and somewhat surprising, being endogenous rather than exogenous. So far, our agents have chosen new neighbors at every step, based on their current interests. What would happen if we assign each agent to a fixed group at the outset, using a fixed similarity threshold?

The behavior depends on the structure of the graph induced by a given threshold. Fig. 6 shows how the number of components depends on the threshold for groups formed in popula-

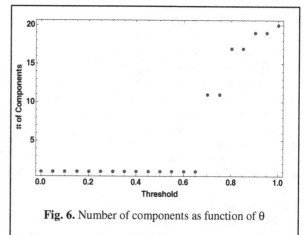

Fig. 6. Number of components as function of θ

tions of 20 agents with 10 interests each. The shift from many components at 0.9 to a few at 0.6 is an instance of the well-known phase transition in random graphs in which a giant connected component emerges as the number of links increases [8], in this case as a result of lowering the threshold. Four cases merit attention.

If θ is very high, there are 20 components, one for each agent. With no neighbors to reinforce its interests, each agent will begin to forget them, and the agents will independently approach the fixed point of an all-zero interest string.

With a low θ, all agents will form one group, and converge as in Fig. 2.

At intermediate θ above the phase shift, each agent's set of neighbors the agents clump into small disjoint components. Each of these groups evolves independently, yielding high diversity among groups but collapse within groups. This model corresponds to allopatric speciation, in which physical separation allows groups to evolve separately.

For intermediate θ below the phase shift, the agents form a number of neighborhoods, but some agents ("bridging agents") belong to more than one neighborhood. Fig. 7 is a graph of one such case with θ = 0.7, with an edge between two agents if the similarity between those agents is greater than θ. Because neighborhoods are fixed over the run, each neighborhood can converge relatively independently of the others, but the bridging agents (in this case, for example, agent 20) repeatedly displace each neighborhood's equilibrium with the emerging equilibrium of another group. Convergence within local

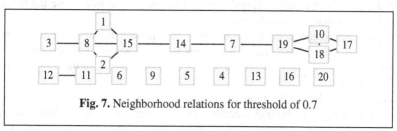

Fig. 7. Neighborhood relations for threshold of 0.7

neighborhoods provides the source of diversity that, mediated by bridging agents, keeps nearby neighborhoods from collapsing. The model of Bednar et al. [3] can be aligned with this result by drawing on their observation that the pressure to internal consistency for a single agent is formally equivalent to the pressure to conformity among a group of agents.

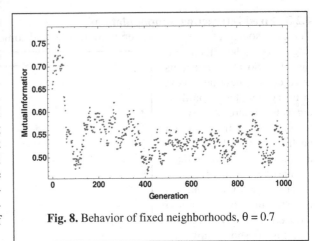

Fig. 8. Behavior of fixed neighborhoods, $\theta = 0.7$

Fig. 8 shows the behavior of this interplay of separate but linked groups. Diversity actually increases over the first 50 generations, as individual groups explore their local configuration of topics. Then, the interaction between groups leads to convergence, but the population does not collapse because each group exerts a pressure toward its own consensus. This mechanism, like curmudgeons and unlike mutation, is robust against intermittent collapse. It reflects a community whose subdisciplines recognize the value of members who bridge with other subdisciplines and exchange ideas between them. Such members are likely to be tolerated better by subgroups than would curmudgeons, because the source of the variation introduced by the bridging individuals is perceived as resulting from their multidisciplinary orientation rather than their orneriness.

This last mechanism is related to Sunstein's observation that polarization is more likely if people feel strong solidarity with their group. By definition, bridging individuals are part of multiple groups. They are less completely identified with a single group, and thus unlikely to be drawn completely into the group consensus. As a result, they can keep the group leavened with new ideas, protecting against collapse.

5   Directions for Future Work

A number of directions for further work suggest themselves. For example:

- We allow agents to have only binary interests. In practice, a person's level of interest in a topic is more nuanced, and it would be interesting to explore the behavior of models with real-valued interests.
- Our topics are independent of one another. In practice, topics may have an influence on one another, which might be represented by a weight between any two topics. Increase in an agent's interest in one topic would lead to an increase in interest in related topics.
- An analytical model of C^3 (compare our previous work on multi-agent convergence [23]) might suggest additional mechanisms for monitoring and avoiding collapse. Existing work on the mathematics of speciation offers a promising foundation,

though our combination of vector beliefs and preferential group interaction is significantly more complex than the systems that have been analyzed previously. A forthcoming paper [21] presents a formal analysis for the case of a single interacting population ($\theta = 0$).

- How can convergence be monitored in practice? Our metric, while effective for simulation, is impractical for monitoring actual groups of people. Explicit questionnaires [26] are appropriate for experimental settings but not for monitoring groups "in the wild." One might monitor the amount of jargon that a group uses, or lack of innovation. A promising example of initial work in this area is Schemer [4].

- Convergence is a two-edged sword. What is degree of convergence allows the production of specialist knowledge without risking collapse?

- How does convergence vary with group size? Recent work [19] suggests that convergence in small groups requires specialized knowledge, while convergence in large groups requires a general knowledge base.

- We have assumed homogeneous tendencies to learn, forget, mutate, or behave curmudgeonly over all agents. How does the system respond if agents vary on these parameters? In particular, what is the impact of these parameters for bridging individuals in comparison with non-bridging individuals?

6 Conclusion

Groups of people naturally converge cognitively. This convergence facilitates mutual understanding and coordination, but if left unchecked can lead to cognitive collapse, blinding the group to other viewpoints.

Experiments with a simple agent-based model show that seemingly obvious mechanisms do not check this tendency. In the domain of academic conferences, these mechanisms represent plenary sessions, special tracks, and random mixing. A source of variation must be introduced to counteract the natural tendency to converge. Mutation is effective if just the right amount is applied, but tends to let the system intermittently collapse. Curmudgeons are more robust, but socially distasteful. Perhaps the most desirable mechanism consists of bridge individuals who provide interaction between individually converging subpopulations. These individuals arise when groups are well-defined, but have thresholds for participation low enough that some individuals can participate in multiple groups.

Insights from this simple model can help monitor and manage collaboration. For example, consider the problem of academic overspecialization. Topical conference tracks can contribute to collapse. The narrow focus of such tracks is enhanced by selecting reviewers for each paper who are experts in the domain of the paper. Papers must be well aligned with the subdiscipline to rank high with such experts, and bridging papers are at a disadvantage. What if one reviewer for each paper were a senior researcher (thus capable of discerning high quality in problem formulation and execution) but *not* a member of the paper's main topic (and thus less disposed to exclude papers that cross disciplinary boundaries)? Such a scheme might encourage the acceptance of quality papers that would otherwise fall in the cracks between subspecialties, and the presence of these papers in topically-organized conference tracks would then provide the bridging function that avoids collapse in our experiments.

References

[1] Andersson, M.: Sexual Selection. Princeton University Press, Princeton (1994)
[2] Axelrod, R.: The Dissemination of Culture: A Model with Local Convergence and Global Polarization. Journal of Conflict Resolution 41(2), 203–226 (1997), http://www-personal.umich.edu/~axe/research/Dissemination.pdf
[3] Bednar, J., Page, S., Bramson, A., Jones-Rooy, A.: Conformity, Consistency, and Cultural Heterogeneity. In: Proceedings of Annual Meeting of the American Political Science Association, Marriott, Loews Philadelphia, and the Pennsylvania Convention Center, Philadelphia, PA, American Political Science Association (2006), http://www.allacademic.com/meta/p150817_index.html
[4] Behrens, C., Shim, H.-S., Bassu, D.: Schemer: Consensus-Based Knowledge Validation and Collaboration Services for Virtual Teams of Intelligence Experts. In: Popp, R.L., Yen, J. (eds.) Emergent information technologies and enabling policies for counterterrorism, pp. 209–230. John Wiley & Sons, Hoboken (2006)
[5] Coyne, J.A., Orr, H.A.: Speciation. Sinauer Associates, Sunderland (2004)
[6] Dawkins, R.: The Selfish Gene. Oxford University Press, Oxford (1976)
[7] Deffuant, G., Neau, D., Amblard, F., Weisbuch, G.: Mixing beliefs among interacting agents. Advances in Complex Systems 3(1-4), 87–98 (2000), http://www.lps.ens.fr/~weisbuch/mixbel.ps
[8] Erdös, P., Rényi, A.: On the Evolution of Random Graphs. Magyar Tud. Akad. Mat. Kutató Int. Közl. 5, 17–61 (1960)
[9] Fisher, R.A.: The Genetical Theory of Natural Selection, p. 193. Clarendon Press, Oxford
[10] Flache, M.W.M.: What sustains cultural diversity and what undermines it? Axelrod and beyond. University of Groningen, Groningen, the Netherlands (2006), http://arxiv.org/abs/physics/0604201
[11] French, J.R.P.: A formal theory of social power. Psychological Review 63, 181–194 (1956)
[12] Futuyma, D.J.: Evolutionary Biology, 3rd edn. Sinauer, Sundarlund (1998)
[13] Hartl, D.L., Clark, A.G.: Principles of Population Genetics, 2nd edn. Sinauer, Sunderland (1989)
[14] Hegselmann, R., Krause, U.: Opinion Dynamics and Bounded Confidence Models, Analysis, and Simulation. Journal of Artifical Societies and Social Simulation 5(3) (2002), http://jasss.soc.surrey.ac.uk/5/3/2.html
[15] Higgs, P.G., Derrida, B.: Stochastic models for species formation in evolving populations. J. Phys. A24, L985–L991 (1991), http://www.lps.ens.fr/~derrida/PAPIERS/1991/higgsspeciation.pdf
[16] Hilscher, R.: Agent-based models of competitive specitationI: effects of mate search tactics and ecological conditions. Evolutionary Ecology Research 7(7), 943–971 (2005)
[17] Krause, U.: A Discrete Nonlinear and Non-autonomous Model of Consensus Formation. In: Elaydi, S., Ladas, G., Popenda, J., Rakowski, J. (eds.) Communications in Difference Equations. Gordon and Breach Science Publ. (2000), http://www.informatik.unibremen.de/~krause/download/ConsensusFormation.pdf
[18] Long, G., Yun-feng, C., Xu, C.: The evolution of opinions on scale-free networks. Frontiers of Physics in China 1(4), 506–509 (2006), http://www.springerlink.com/content/74x0w76817014272/

[19] Palla, G., Barabási, A.-L., Vicsek, T.: Quantifying social group evolution. Nature 446, 664–667 (2006), http://arxiv.org/abs/0704.0744

[20] Palla, G., Barabási, A.-L., Vicsek, T.: Quantifying social group evolution. Nature 446(7136), 664–667 (2007),
http://www.nd.edu/~alb/Publication06/
142-SocialGroup_Nature-5Apr07-CN/SocialGroup_Nature5Apr07.pdf

[21] Parunak, H.V.D.: A Mathematical Analysis of Collective Cognitive Convergence. In: Proceedings of the Eighth International Conference on Autonomous Agents and Multi-Agent Systems (AAMAS 2009), Budapest, Hungary (submitted, 2009),
http://www.newvectors.net/staff/parunakv/AAMAS09CCC.pdf

[22] Parunak, H.V.D., Belding, T.C., Hilscher, R., Brueckner, S.: Modeling and Managing Collective Cognitive Convergence. In: Proceedings of The Seventh International Conference on Autonomous Agents and Multi-Agent Systems. International Foundation for Autonomous Agents and Multi-Agent Systems, Estoril, Portugal, pp. 1505–1508 (2008),
http://www.newvectors.net/staff/parunakv/AAMAS08M2C3.pdf

[23] Parunak, H.V.D., Brueckner, S.A., Sauter, J.A., Matthews, R.: Global Convergence of Local Agent Behaviors. In: Proceedings of Fourth International Joint Conference on Autonomous Agents and Multi-Agent Systems (AAMAS 2005), Utrecht, The Netherlands, pp. 305–312. ACM, New York (2005),
http://www.newvectors.net/staff/parunakv/AAMAS05Converge.pdf

[24] Pennings, P.S., Kopp, M., Meszéna, G., Dieckmann, U., Hermisson, J.: An Analytically Tractable Model for Competitive Speciation. The American Naturalist 171(1), E44–E71 (2008)

[25] Provine, W.B.: Sewall Wright and Evolutionary Biology. University of Chicago Press, Chicago (1986)

[26] Romney, A.K., Weller, S.C., Batchelder, W.H.: Culture as Consensus: A Theory of Culture and Informant Accuracy. American Anthropologist 88(2), 313–338 (1986)

[27] Sunstein, C.R.: The Law of Group Polarization. 91, University of Chicago Law School, Chicago, IL (1999), http://ssrn.com/paper=199668

[28] Sznajd-Weron, K., Sznajd, J.O.: Opinion Evolution in Closed Community. International Journal of Modern Physics C 11(6), 1157–1165 (2000),
http://www.ift.uni.wroc.pl/~kweron/articles/
Sznajd-WeronSznajd00_IJMPC.pdf

Dynamics of Agent Organizations: Application to Modeling Irregular Warfare

Maksim Tsvetovat and Maciej Latek*

Center for Social Complexity,
George Mason University,
4400 University Drive, Fairfax VA 22030
{mlatek,mtsvetov}@gmu.edu
http://www.css.gmu.edu

Abstract. In this paper, we focus on computational modeling of adversarial activities and asymmetric warfare in a tactical setting. As a topic for simulation study, asymmetric warfare is an odd and ill-conditioned problem. Empirical data is scarce or one-sided, while the subject itself presents a constantly adapting and moving target that makes it a strategic priority to conceal its inner workings from the observers.

To provide an insight into the dynamics of the asymmetric conflicts, one needs to constrain the model in rigorous social-scientific concepts, including those of organization theory, theory of collective action, and social network analysis.

Our model, called NetMason, enables controlled experiments to replicate and analyze alternative policies of disruption of activities of terrorist organizations. In addition, sensitivity analysis with respect to behavioral assumptions can be easily performed.

Keywords: Multiagent Simulation, Organizational Learning, Social Network Analysis, Dynamic Social Networks, Terrorist Networks, Irregular Warfare.

1 Introduction

As a subject for simulation study, asymmetric warfare is an odd and ill-conditioned problem. Empirical data in this field is scarce or one-sided, while the subject itself presents a constantly adapting and moving target, that makes it a strategic priority to conceal its inner workings from the observers, [10]. Insurgencies evolve in step with counter-insurgency efforts, resulting in a co-evolutionary process. Terrorist organizations do not exist in a political or social vacuum [8], and complex logistics of the terrorist operations are just as important as motivations and resources available to the insurgency.

Social network analysis (SNA) is acknowledged as a powerful tool in this context, see [6]. Recently, a SNA model of dynamic terrorist network has been

---

* This work is sponsored in part by the Center for Social Complexity at George Mason University. Opinions expressed herein are solely of the authors and are not these of George Mason University.

N. David and J.S. Sichmann (Eds.): MABS 2008, LNAI 5269, pp. 60–70, 2009.

proposed, namely a *cellular network* [15,19,11]. A cellular social network is a non-traditional organizational configuration consisting of quasi-independent cells of densely connected agents and a distributed chain of command connecting cells in a non-hierarchical fashion.

Despite the progress in SNA research, there is a significant and growing gap between the desired counterinsurgency analysis tools needed and the available and relevant computational social science. Traditional SNA models lack the purposeful agents and can not account for organizational dynamics and evolution. At the same time, a number of agent-based solutions have been proposed to model organizational change. For example, K. Carley [5,3] has proposed a model (DyNet) to account for organizational processes such as information diffusion, institutionalization and recovery of organizations from disruptions. While DyNet can account for macroscopic properties of the organizational evolution, its micro-level foundations are based on fixed-strategy agents.

In this paper, we demonstrate a robust model of operations of terrorist organizations. Our model, called NetMason, combines the ability to simulate complex, large, and decentralized organizational networks with the ability to realistically represent individual agents and interactions. Such a realism is enabled by the use of artificial intelligence (AI) techniques and robust knowledge representation — as a backbone of every agent.

NetMason is intended to bring quantitative reasoning into the field of irregular warfare modeling by enabling controlled, computational experiments. These experiments can be used to replicate and analyze alternative policies in terms of their impact on the daily functioning of the insurgency and levels of violence.

2   NetMason Architecture

We structured the mental model for agents according to Belief-Desire-Intention (BDI) architecture, see [14]. In BDI structures, agents differentiate between their perception of the world, long-term strategic goals and short-term action plans and use planning algorithms to tie these concepts into sequences of behaviors and actions. This approach has been widely applied in the AI community, and has found implications in modeling of dynamic organizational networks, for example [22,1].

One could argue that in human organizations, the variability in decision-making originates from two sources. The first source is the differentiated positions on the social network and access to resources and knowledge. The second is the agents' inherent heterogeneity in decision-making abilities. It has been a commonly raised argument against BDI architectures for simulation of organizational systems that BDI imposes a homogeneous behavioral logic amongst agents, see [1]. We avoid this problem by extending the BDI backbone and allowing not only differentiation in agents' roles and access to information. In NetMason, details of planning and communication procedures can be customized for each of the agents using a library of behavioral scripts.

Consequently, each of agents is a set of independent components joined by a common interface (see Figure 1). Central to the dynamics of an agent are

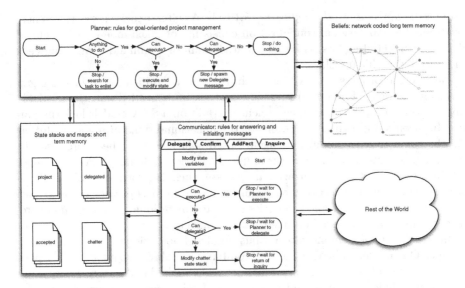

Fig. 1. Graphic shows the main components of each individual agent, with flowcharts representing scripts used to model the 1998 bombing of the US Embassy in Dar es Salaam. Each of the components can be customized with agent-specific Groovy scripts that can be specified in the scenario file.

Planner and Communicator modules, which are responsible for invoking planning and communications scripts. The BDI architecture requires separation of agent knowledge and perception of the world into short-term (intentions) and long-term (beliefs) components:

- **Intentions.** Intentions are represented by a set of maps and stacks that preserve information about current actions of an agent and it's strategic goals, necessary to implement Hierarchical Task Network (HTN) planning procedures and task timeouts. The project plans representing procedural constraints for HTN are common for all agents and are part of the environment.
- **Beliefs.** Beliefs of an agent are represented as a network of facts. This network is used to store information about agent's information about contacts, resources and capabilities of it's peers. This network also encodes the agent's transactive memory, see [21], storing acquired interaction histories as edge properties. Each of the elements of the network can be labeled with a confidence value.

The Communicator module has two basic functions. First, it is responsible for parsing of incoming messages. Second, it includes procedures for self-initiated chatter and for controlling the trust and confidence into incoming traffic. Chatter is an essential social activity undertaken by agents. It allows informal networks and weak ties between agents to form, and facilitates informal transfer of knowledge.

Currently, agent must know how to interpret and respond to following classes of messages:

- **InquireAbout, AddFact.** When an agent determines that a resource, capability or information absent from it's long-term memory is required to proceed with the project, InquireAbout message will be initiated and send to a selected peer. Respondent might answers with AddFact message that contains relevant piece of knowledge.
- **TakeTask, ConfirmActivity, DelegateActivity.** When an agent is convinced some other agent is able to execute a task, TakeTask message might be send. The contractor reportes successful executions or failures.
- **MeetAgent, TakeResource.** Instead of executing a task itself, an agent might choose to introduce delegating agent to potential contractor using MeetAgent message. Similarly, transfer of resource over the social network is accomplished with TakeResource message.

We use a minimal subset of messages required for the organization to accomplish projects, but this can be easily extended with addition of new behavioral scripts. The choice of communication partners, selection of messages and responses and all memory-based calculations are subject to direct trust principles of [9].

The Planner module uses derivative of a Hierarchical Task Network decomposition algorithm to service tasks represented using task precedence and dependence networks, as in [18,17]. In the simulated scenarios constraints among tasks, resources and capabilities are described in the form of so called task network. Given target task, Planner will proceed and recursively solve for the sequence of other tasks that need to be executed and resources and capabilities to acquire in order to fulfill all of the goal's dependencies.

Planner module exploits the fact that every agent is embedded in an organization when trying to obtain the goals. Using default behaviors, the Planner may initiate messages querying for help and delegate activities to other agents. For example, if required task dependency is not found in long-term memory, InquireAbout message will be send. If agent who is able to execute the task or has access to dependency is present in memory, message TakeTask might be send.

The last element of the BDI architecture, desires (objectives to be accomplished), are exogenous to the agents. In the current version of NetMason agents have no mechanism for self-initiating projects. Therefore, initial matching between agents and high-level tasks is provided in the simulated scenarios. Afterwards, this matching evolves through delegating the tasks subject to trust constraints and changing social network.

3   Using NetMason

To instantiate the ideas from previous section, we joined together a number of open source solutions. The core of current version of NetMason is the MASON simulation framework that provides a fast discrete-event multi-agent simulation library core in Java, presented in [12]. For efficient representation of beliefs and basic graph operations, we use the JUNG social network package, see [13]. The database persistence and required experimental instrumentation is provided by Apache Cayenne persistence framework. Lastly, to increase scenario flexibility we use the Mozilla Rhino scripting engine. Scripting enables users to change the behavioral logic of agents or create new types of interactions without need for recompiling the framework. We plan that as NetMason evolves, future models will be built by using collections of scripts rather then by changes to the model infrastructure itself.

The simulation is started by loading a scenario from an XML file using DyNetML, see [2], see [20]. In particular, a scenario file contains information on:

1. Initial social network, used to generate initial beliefs for agents;
2. Initial matching between agents and capabilities / resources;
3. Specific behavioral scripts to be used by each of the agents;
4. Pre-generated plan templates specifying task hierarchies and resource / capabilities dependencies for each of the subtasks;
5. Pre-scheduled events, including initial task assignments.

Each simulation run, including details of all events and messages, is stored in a relational database using the persistence layer. The database is consistent with DyNetML and can be used to feed data into reporting, statistical or social network analysis software. This allows to use existing tools like ORA to visualize and analyze the evolution of terrorist organizations.

One of the MASON features that is being exploited is strict separation of visualization and simulations layers. This feature allows for the same version of our model to be run efficiently on a single core desktop machine with all GUI features turned on or in a batch mode distributed over a Linux cluster of high end multi-core nodes feeding results into a common database[1].

4   Experiments

This section will feature a practical application of NetMason. It's purpose is to demonstrate the current state of NetMason and establish technical and process validity of selected algorithms. In order to facilitate this process we have created a set of test scenarios, one of which is described in the next Subsection in more detail.

4.1   Dar-es-Salaam Scenario

The scenario is based on data gathered in the wake of the 1998 bombing of the US Embassy in Dar-es-Salaam. It is composed of following elements:

- **Social network.** Network consists of 14 terrorists with differentiated access to resources (four different resources are bomb material, money, truck and bomb tooling) and knowledge (bomb building expertise, weapons specialist, surveillance expertise).

---

[1] MASON was the first ABM system to provide this key feature in computational social science and remains the leading system in this respect.

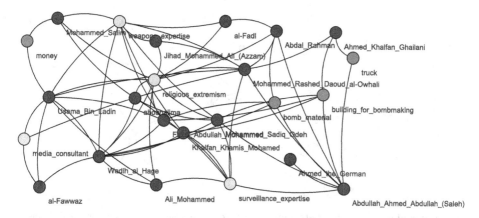

Fig. 2. View of the Dar-es-Salaam scenario file, with task structure removed. For clearer view of underlying agent×agent network, please refer to Figure 4. Legend: ● agents, ● resources, ● capabilities.

- **Task structure.** The project they are tasked with (bombing) consists of two smaller sub projects (building the bomb itself and performing target surveillance) in addition to securing the transportation and financial resources. All in all, the task hierarchy contains three tasks with nine preconditions.
- **Dynamic structure.** This includes names of behavioral scripts used for each agent as well "deus ex machina" messages that will be used to trigger certain events (like initial task assignment in default experiment or death / removal of an agent as used in later experiments).

Figure 2 presents the full initial social network and capability network for the Dar-es-Salaam scenario. For clearer view of the initial social network, please refer to Figure 4. The Dar-es-Salaam scenario is used in three separate experiments that are intended to present different aspects of NetMason dynamics and validate it against the extant body of organizational learning literature.

4.2 Micro-level Validation

The first set of experiments explores NetMason's execution dynamics and their accordance with assumptions laid out in the previous Section:

Experiment 1. Evolution of beliefs of agents and communications taking place during a simulation run will be visualized. In this experiment, default configuration of the scenario will be used and a chosen agents (Usama bin Laden) will be asked to execute a single bombing.

Figures 3 and 4 show evolution of beliefs for a sample agent as well as predicted intensity of communication taking place, reprojected onto the initial social network.

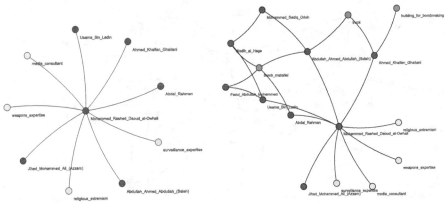

(a) Initial configuration of beliefs. Only first-degree facts (certanities) read of the scenario file are present.

(b) Terminal configuration after project has been executed. Agent has learned about other agents and their properties.

Fig. 3. Evolution of long-term memory (beliefs) for agent Mohammed Rashed Daoud. Prepared for Experiment 1. Legend: ● agents, ● resources, ● capabilities.

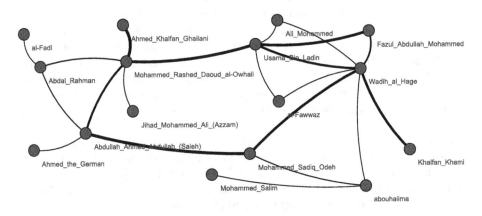

Fig. 4. Agent to agent communication intensity recorded during a single run of Experiment 1. The link width is proportional to the total number of messages passed between a pair of agents.

Experiment 2. Improvement in team performance as a function of task repetitions will be assessed. As in the default scenario, group is always able accomplish the bombing task, therefore as a metric of interest, time to bombing completion will be used.

Figure 5 shows distribution of times necessary to gather all dependencies and finalize the execution of three subsequent bombings. A significant decrease in the total execution time is observed after successful accomplishment of the first bombing. This is due to the establishment of trusted pathways in the organization,

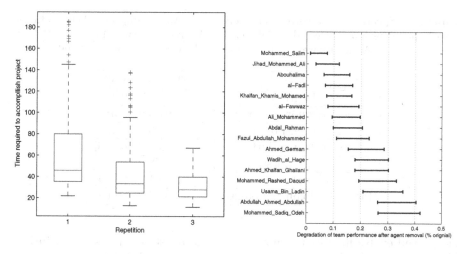

Fig. 5. Does organization's performance improve with repetition? Graph presents distribution of time required to perform a single bombing attack as a function of the organization's experience. Legend: + outliers, ⊥ min-max range, ▯ inter-quartile range, — median. Prepared using Experiment 2 setup.

Fig. 6. How does elimination of an agent affects the performance of the terrorist team? Shown is decrease in proportion of accomplished tasks after agent is removed, with associated 95% confidence intervals. Calculated using Experiment 3 setup.

necessary to gather different project components in an efficient manner. Organizational learning during the second and third tasks affects performance of the group in a marginal way, demonstrating the law of diminishing returns. In Experiment 2, 300 repetitions per scenario modification (additional task, agent removal) were performed and the results used to calculate desired statistics. Following Sections contain detailed discussion of results.

4.3   Macro-level Validation

The goal of the next experiment is to generate macro-level statistics of performance of the terrorist team under varying conditions:

> **Experiment 3.** Robustness of terrorist group with respect to the elimination of some of the members will be studied. A metrics of interest will be the proportion of tasks completed as a function of terrorist eliminated. Those values will be compared to predictions of Organizational Risk Analyzer [4].

Experiment 3 explores the one of the pivotal questions in the security-related analytics, namely the issue of discovering the optimal method of disrupting the activities of a terrorist team when only partial information is available. Currently, such problems are solved by application of SNA heuristics. It should be noted, that their static nature can not a account for the fact that the terrorist networks

Table 1. R-Spearman rank correlations between NetMason Impact indicies obtained from Experiment 3 and selected Social Network Analysis measures for Dar-es-Salaam scenario. The SNA metrics were computed using the Organization Risk Analyzer [4].

Metric	Spearman-R	p-value
Degree	0.215	0.422
Socio Economic Power	0.235	0.380
Bonacich Power	0.385	0.139
Constraint Burt	-0.553	0.026
Betweenness	0.589	0.016
Capability	0.614	0.004
Boundary Spanner Potential	0.674	0.002

morph and evolve fluidly in response to anti-terrorist activity, see [16]. In Net-Mason, after setting initial conditions, the organizational structure of a terrorist network is not fixed, instead it dynamically emerges from the local constraints that require maintenance of secrecy (via the trust heuristics) to be balanced with operational efficiency. NetMason provides a dynamic solution to the disruption problem. Thus, estimates of the effect of the removal of an individual on the operations of a particular cell can be regarded as more dependable than those obtained from classical SNA.

For each agent removed, decrease in the long-term performance of the whole team was recorded. By repeating this scheme 300 times for each of the agents, we obtain estimates of how many of tasks can be accomplished within allotted time as compared to how many the full team would accomplish as a function of agent / group of agents removed. Average value of this measure for a given agent will be later referred to as NetMason Impact. This way we obtain a clear ranking of agents that can be potentially targeted for removal, presented on Figure 6.

In Table 1 we correlate NetMason Impact with SNA recommendations, observing a number of high correlations. From all of the SNA measures available in ORA, three correlate particularly high with NetMason Impact: Betweenness, Capability and Boundary Spanner Potential. Betweenees emphasizes role of an agent in information transfer, Capability in access to knowledge and resources and Boundary Spanner Potential accounts for importance of agent in unifying sub communities of organization and contributing to robustness of it's structure. NetMason Impact integrates those measures into a scalar value, at the same time accounting for dynamics of a particular scenario.

5   Summary

This paper presents the first version of NetMason, a computational model for analysis of irregular warfare. NetMason combines specific knowledge about irregular conflicts (expressed using scenario files) with social network analysis, agent-based modeling, automated event data analysis and other supporting modeling technologies. General outline of modeling philosophy as well as results of very first validation experiments were also included.

Experiments 1 and 2 demonstrated face validity of NetMason, both on organizational (macro) level as well as on the level of individual dynamics (micro). Experiment 3 docked our model within the extant SNA results. To clearly evidence value-added NetMason offers, more elaborate validation study is necessary that would use new datasets.

NetMason is currently in active development. Apart from enhancing the scenario library, our goals for the nearest future include following augmentations:

Agent Strategic Reasoning. NetMason agents should be able to make strategic decisions and choose for themselves high-level goals (desires), as well as communicate them to other agents in the network. This will allow for simulating the spread of insurgency and tactics it uses. We shall utilize a probabilistic mechanism of decision-making, informing the prior probabilities from empirical data collected during investigations of terrorist incidents.

Friend-or-Foe perception. While terrorist groups cause significant harm and loss of life to local populations, they can be perceived as "freedom fighters" or friendly forces, while counter-insurgency (COIN) operations are viewed as enemy raids. If such mentality persists, the counter-insurgency forces cannot achieve their mission and face increasing militancy in the population. NetMason should model evolution of friend-or-foe attitudes among the population, and includes facilities for *in-silico* experiments for influencing these attitudes based on social science theories of cognitive balance and norm adoption [7].

Logistics and Economy. As mentioned earlier, insurgent and terrorist activities often result in the creation of underground economies and cottage industries for providing supplies for covert organizations. We aim to represent such feedbacks, with focus on the market for building, distribution, and operation of Improvised Explosive Devices (IED's). This will be accounted for by allowing agents to strategically manage and exchange resources (e.g. weapons, money) and their time and labor.

References

1. Canuto, A.M.P., Campos, A.M.C., Santos, A.M., Moura, E.C.M., Santos, E.B., Soares, R.G., Dantas, K.A.A.: Simulating working environments through the use of personality-based agents. In: Sichman, J.S., Coelho, H., Rezende, S.O. (eds.) IBERAMIA 2006 and SBIA 2006. LNCS, vol. 4140, pp. 108–117. Springer, Heidelberg (2006)
2. Carley, K.M.: Smart agents and organizations of the future. In: Lievrouw, L., Livingstone, S. (eds.) The Handbook of New Media, ch. 12, pp. 206–220. Sage, Thousand Oaks (2002)
3. Carley, K.M.: Dynamic network analysis. In: Breiger, K.C.R., Pattison, P. (eds.) Dynamic Social Network Modeling and Analysis: Workshop Summary and Papers, pp. 361–370. Committee on Human Factors, National Research Council (2003)
4. Carley, K.M., Reminga, J.: Ora: Organization risk analyzer. Technical Report Technical Report CMU-ISRI-04-101, Carnegie Mellon University, School of Computer Science, Institute for Software Research International (2004)

5. Carley, K.M., Reminga, J., Borgatti, S.: Destabilizing dynamic networks under conditions of uncertainty. In: IEEE KIMAS, Boston, MA (2003)
6. Falkenrath, R.: Analytic models and policy prescription: Understanding recent innovation in counterterrorism. Technical Report BCSIA Discussion Paper 2000-31, John F. Kennedy School of Government, Harvard University (2000)
7. Galan, J., Latek, M., Tsvetovat, M., Rizi, S.: Axelrod's metanorm games on complex networks. In: AGENT 2007 (2007)
8. Gerwehr, S., Daly, S.: Al-Quaida: Terrorist Selection and Recruitment, ch. 5, pp. 73–89. Rand Corporation (2005)
9. Gomez, M., Carbo, J., Earle, C.B.: Honesty and trust revisited: the advantages of being neutral about others cognitive models. Auton. Agent Multi-Agent Syst. 15, 313–335 (2007)
10. Ilachinski, A.: Artificial War: Multi-Agent Based Simulation of Combat. World Scientific, Singapore (2004)
11. Lawrence, P., Lorsch, J.: Differentiation and integration in complex organizations. Administrative Science Quarterly (12), 1–47 (1967)
12. Luke, S., Cioffi-Revilla, C., Panait, L., Sullivan, K., Balan, G.C.: Mason: A multi-agent simulation environment. Simulation 81(7), 517–527 (2005)
13. O'Madadhain, J., Fisher, D., White, S., Boey, Y.-B.: The JUNG (Java Universal Network/Graph) Framework. Technical report, School of Information and Computer Science, University of California, Irvine (2006)
14. Rao, A.S., Georgeff, M.P.: Modeling rational agents within a BDI-architecture. In: Allen, J., Fikes, R., Sandewall, E. (eds.) Proceedings of the 2nd International Conference on Principles of Knowledge Representation and Reasoning (KR 1991), pp. 473–484. Morgan Kaufmann publishers Inc., San Mateo (1991)
15. Rothenberg, R.: From whole cloth: Making up the terrorist netwrok. Connections 24(3), 36–42 (2002)
16. Sageman, M.: Understanding Terror Networks. University of Pennsylvania Press (2004)
17. Tate, A.: Generating project networks. In: Proceedings of IJCAI 1977, pp. 888–893 (1977)
18. Thomson, J.: Organizations in Action. McGraw-Hill, New York (1967)
19. Tsvetovat, M., Carley, K.M.: Simulation of human systems requires multiple levels of complexity. In: Deguchi, H. (ed.) Proceedings of World Congress on Social Simulation, vol. 1 (2006)
20. Tsvetovat, M., Reminga, J., Carley, K.M.: Dynetml: Interchange format for rich social network data. Technical Report CMU-ISRI-04-105, Carnegie Mellon University (2004)
21. Wegner, D.M.: Transactive memory: A contemporary analysis of the group mind. In: Mullen, B., Goethals, G.R. (eds.) Theories of group behavior, pp. 185–208 (1987)
22. Ye, M., Carley, K.: Radar-soar: Towards an artificial orgnization composed of intelligent agents 20(2-3), 219–246 (1995)

Using Simulation to Evaluate Data-Driven Agents

Elizabeth Sklar[1,2] and Ilknur Icke[2]

[1] Dept of Computer and Information Science,
Brooklyn College, City University of New York
2900 Bedford Ave, Brooklyn, NY 11210 USA
sklar@sci.brooklyn.cuny.edu
[2] Dept of Computer Science,
The Graduate Center, City University of New York
365 Fifth Avenue, New York, NY 10016 USA
iicke@gc.cuny.edu

Abstract. We use simulation to evaluate agents derived from humans interacting in a structured on-line environment. The data set was gathered from student users of an adaptive educational assessment. These data illustrate human behavior patterns within the environment, and we employed these data to train agents to emulate these patterns. The goal is to provide a technique for deriving a set of agents from such data, where individual agents emulate particular characteristics of separable groups of human users and the set of agents collectively represents the whole. The work presented here focuses on finding separable groups of human users according to their behavior patterns, and agents are trained to embody the group's behavior. The burden of creating a meaningful training set is shared across a number of users instead of relying on a single user to produce enough data to train an agent. This methodology also effectively smooths out spurious behavior patterns found in individual humans and single performances, resulting in an agent that is a reliable representative of the group's collective behavior. Our demonstrated approach takes data from hundreds of students, learns appropriate groupings of these students and produces agents which we evaluate in a simulated environment. We present details and results of these processes.

1 Introduction

Most work that lies at the intersection of education technology and agent-based systems employs agents within intelligent tutoring systems as knowledgeable, automated teachers. Other work has explored the notion of simulated students [1,2,3] as a means to better understand the processes that underpin human learning by constructing models based on theories from pedagogical and/or cognitive science literature. Some of our earlier work deployed simple agents as learning peers in simple games, where the agent controllers were built from data collected in previously played games [4]. The current work extends these ideas by using

N. David and J.S. Sichmann (Eds.): MABS 2008, LNAI 5269, pp. 71–84, 2009.
© Springer-Verlag Berlin Heidelberg 2009

data from an on-line educational assessment environment to train agents, and we employ simulation as a means for evaluating the agents we derive.

We are not the first to suggest constructing agents to emulate human behaviors in on-line systems [5,6,7]. Previous work has shown that training agents to emulate humans produces better results if the training set is a composite of multiple humans grouped according to application-specific metrics [8]. A large data set is generally desirable when training agents, and grouping human data sets with similar characteristics helps smooth out anomalies. In the study presented here, we apply data-mining techniques to interaction logs from a dynamic, on-line educational assessment environment in order to find effective groupings within the human data set. We then use the data groups to train agents that can be deployed in a simulation environment where each agent effectively embodies a *user model* that is characteristic of the humans in its training group. Related work has been reported previously (e.g., [9,10,11]), but within different user modeling contexts and not with the purpose of subsequently using the models to control agents in a simulation environment.

Our procedure is as follows. The first step is to partition the interaction data set that will serve as the basis for training a suite of distinct, agent-based, probabilistic controllers. This step is crucial to the success of each agent as an emulator of a specific class of human behaviors. If the data used to train the agents is not well defined, then there may be two problems. First, the behavior of individual agents may have too much variation and so an agent would be an unreliable model of the human users in its group. Second, the agents in the suite may not be distinct enough from each other (i.e., their behavior patterns may overlap significantly) and so an agent would be unreliable as a model of only the users in its group. In this paper, we compare different techniques for partitioning the data, using standard clustering methods from the data mining literature. The second step is to produce an agent for each cluster whose actions typify behaviors characteristic of members of the cluster. The final step is to evaluate the results by placing all the agents we generated in a simulated assessment environment and by measuring which agents most closely resemble the human counterparts they are meant to be simulating. Our results demonstrate that one clustering technique clearly produces a better approximation of group behavior patterns than the others, as evaluated through simulation.

2   Our Approach

Figure 1 contains a graph illustrating a generic landscape for an interactive environment. Imagine that each node represents a *state* in the environment, such as a page in a web site, a question in an on-line test or a room in a computer game. The red and green links between nodes indicate paths that users can follow based on actions they take in the interactive environment. For example, clicking on a hyperlink will send a user to another web page, or answering a question correctly will send a user to another question, or earning game points will send a user to another room. All users start their interactions with the system in the

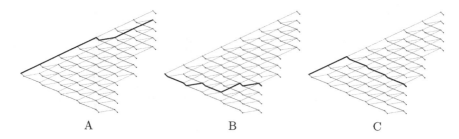

Fig. 1. Sample trajectories

same place—at the node on the far left. A sequence of user actions translates into a *trajectory* through the landscape. Three sample trajectories are drawn with heavy black lines in the figure.

In fact, the figure is modeled after an educational assessment environment. Each node represents a question in the assessment, and links indicate possible paths from one node to another. All students start at the same node, the leftmost one in the figure. Students who answer a question correctly move along the green link (up and to the right). Students who answer a question incorrectly move along the short red link (down and to the right). Thus Figure 1 illustrates the paths taken by three different students. Student A (left side of the figure) did well, answering most questions correctly. Student B (middle of the figure) did poorly at first, rallied, slumped, rallied, and slumped again to the end of the assessment. Student C (right side of the figure) started off well, but then made a series of mistakes. The differences in performance of these three students is clearly visible. Notice in particular that while students B and C ended up at the same node at the end of the assessment, they took very different paths through the landscape.

Most assessments report results and group students using the final, or "exit", score achieved. In our example, this corresponds to the last node visited, on the right edge of the landscape. However, with the advent of dynamic, on-line testing, a much richer data set is available and so reports can provide more information about student performance than simply an exit score. We believe that in environments like the one that corresponds to our sample data set, a lot can be learned from examining the trajectories taken through the landscape. So, we focus on trajectories and experiment with techniques for grouping students according to similarities in their trajectories. It is important to note that the students do not make directed choices about which paths to take, but rather the system chooses each next node in response to the student's performance up to that point. As mentioned earlier, we wish to create a suite of agents that each mimic certain classes of human behavior. Section 2.1 describes several methods for partitioning the complete data set into clusters, grouping humans with similar behavioral characteristics, as exhibited by following similar trajectories through the landscape. The experiments conducted here are based on a data set of 117 students who had accessed the assessment environment in 2006.

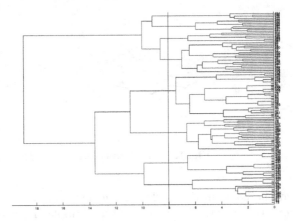

Fig. 2. Hierarchical clustering of student response vectors, using Euclidean distance with 012 coding. The vertical axis identifies individual students. The horizontal axis contains h values.

2.1 Partitioning Training Data

Data clustering is a well-studied field in the literature, and the particular algorithm chosen for a clustering task varies depending on the characteristics of the data set and the goals of the task. Our aim is to produce a coherent grouping that will serve to train a suite of agents that are distinct from each other. We investigated a variety of techniques and here we compare two types: Euclidean distance based on feature vectors and Hausdorff distance based on geometric similarity.

Euclidean distance. We generated "feature vectors" to encode the student responses in the assessment. For the landscape illustrated in Figure 1, we generated 94-dimensional feature vectors for each student, each dimension representing a student's response to one question. Note that sequences of questions are chosen for students dynamically, based on the their performance and the connections defined in the landscape; so students do not (and can not) visit every node. There are three possible results for any question: *correct*, *incorrect* or *not seen*. We experimented with different ways to encode results including: 0=incorrect, 2=correct, 1=not seen (021 coding), −1=incorrect, 1=correct, 0=not seen (-110 coding), 0 =incorrect, 1=correct, 2=not seen (012 coding), and also a 3-variable coding: seen ($\{1|0\}$), incorrect($\{1|0\}$), correct ($\{1|0\}$). We employed a hierarchical clustering algorithm in Matlab [12] using the Euclidean distance between the feature vectors as the distance measure. Note that according to the metrics described in the remainder of this paper, the feature vector encoding that produced the best results is the 012 coding. Thus, for the sake of brevity, we only present the Euclidean 012 coding results here.

Hausdorff distance. Since our aim is to group students according to the similarity of trajectories followed on the assessment maps, we decided to employ

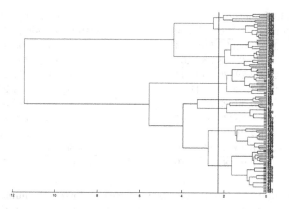

Fig. 3. Hierarchical clustering of student trajectories, using Hausdorff distance. The vertical axis identifies individual students. The horizontal axis contains h values.

clustering techniques on sequential data. Similar work has been reported on classifying Linux users with respect to their experience level based on command logs [13] and on clustering financial time series data [14]. We generated the landscape illustrated in Figure 1 by assigning coordinates to each question in the educational assessment, and we used the Hausdorff Distance to compute the dissimilarity between any two paths. After computing the pairwise distances between each path, we applied a hierarchical clustering algorithm in Matlab.

2.2   Comparing Clusterings

We compare the results of different clustering algorithms by measuring the distance between data points within each cluster and the separation between clusters, seeking to minimize *intra*-cluster differences (the level of similarity within a cluster) and maximize *inter*-cluster differences (the amount of dissimilarity from one cluster to another).

One way of comparing clustering results is using a type of figure called a *dendrogram*. A dendrogram consists of brackets that connect objects hierarchically. The height (h) of each bracket indicates the distance between any two objects (or groups of objects) being connected. Figures 2 and 3 show dendrograms for the Euclidean 012 coding and Hausdorff methods, respectively. The Euclidean 012 method found 8 main clusters at level $h = 8$ while the Hausdorff method found 8 main clusters at level $h = 2$. Because lower h values indicate less difference amongst group members, the Hausdorff method has better results.

Another way we compare the clustering results is with metrics that have been demonstrated to compare clusters of trajectories in related work [15]. Two metrics are employed. The first is "shape complexity", or σ, which is computed as:

$$\sigma = disp/length$$

where $disp$ is the displacement or the distance between the first and last points in a trajectory and $length$ is the number of points in the trajectory. The second is

Hausdorff method				
cluster	size	points	σ	cov
1	11	121	2.64025	5687.83
2	6	64	2.42758	3871.61
3	19	203	2.22529	6487.44
4	28	298	4.43650	6841.44
5	4	37	4.89523	6909.63
6	12	108	4.55438	5109.01
7	18	197	4.91468	8704.28
8	19	206	5.40067	8579.10
average			3.93682	6523.79

Euclidean 012 method				
cluster	size	points	σ	cov
1	6	68	4.87123	8353.60
2	19	204	8.83936	15166.20
3	4	41	2.20674	4062.49
4	8	89	2.14947	7424.48
5	16	175	3.29415	4157.30
6	15	159	1.60082	4469.70
7	21	223	7.63581	12024.50
8	28	275	7.57357	7320.25
average			4.77139	7872.31

Fig. 4. Cluster similarity measures, showing for each cluster: the number of trajectories in the cluster, the total number of points covered by all the trajectories in the cluster, the standard deviation for σ and the standard devation for cov (see text for further explanation)

"divergence", or covariance of the first, middle and last points in the trajectory. Figure 4 compares these values for the Euclidean 012 and Hausdorff methods. The size of each cluster (number of students belonging) is shown as well as the average number of points in the trajectories of all student members. The columns to focus on are the two rightmost, which contain the standard deviation of σ and cov for the trajectories that comprise each cluster. The absolute numbers are not important here; what is important is the relationship between the numbers within each column. Smaller numbers indicate tighter coherence amongst cluster members—this is our aim. The average σ for the Hausdorff is 3.95, whereas for Euclidean, the average $\sigma = 4.77$. The average covariance for Hausdorff is 6523.79, whereas for Euclidean, the average cov $= 7872.31$. Using these metrics, the Hausdorff distance clustering technique results in better coherence. We note that a cluster-by-cluster comparison reveals that the Hausdorff coherence is better for the clusters on either end of the spectrum, while the Euclidean is better for those in the middle. This is an interesting result which bears further investigation.

One additional factor to take into account with the clustering methods is the choice of number of clusters. The more clusters there are, the better the coherence amongst cluster members (and the smaller the standard deviation of our two metrics). However, there is a trade-off: perfect coherence can be achieved with clusters of size 1, but being satisified with single-member clusters implies that clustering is not needed at all. Thus, a balance must be achieved between the number of clusters and the coherence. Figure 5 illustrates the change σ and cov as the number of clusters decreases. The values in the figure were computed for the Hausdorff methodology (which will be shown later to be the best overall choice of clustering algorithm for our problem). For both metrics, when the number of clusters reaches 5 and again at 10, the angle of decline becomes less sharp. To allow for fair comparison between algorithms, we chose a standard number of clusters for each algorithm: 8.

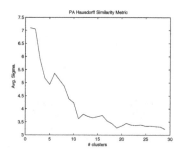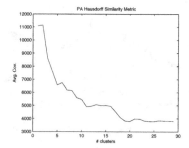

Fig. 5. Change in σ and *div* as number of clusters decreases (Hausdorff clustering algorithm)

Recall that our overall goal is to generate clusters that will provide training sets for a suite of agents, where each agent effectively embodies a user model that is characteristic of the humans in its training group. Section 2.3 describes how we trained agents based on the clustered data. Section 2.4 presents evaluation results that demonstrate the effectiveness of the clustering and training procedure.

2.3   Training Agents

The next step in our procedure is to create agents whose behavior typifies that of each cluster. We did this for clusters generated by the Hausdorff and all Euclidean coding methods (though we only report here on the best coding, 012). First, we generated a profile for each cluster as follows. For each node in the landscape, we tally the number of students in the cluster who visited that node. We compute statistics based on students' responses to each node as the basis, aggregated for all members of a cluster into an agent training set. For each cluster, we generate one representative agent.

Each node represents a question in the assessment, and each question is designed to elicit information about students' abilities. The assessment is a multiple-choice test, and each possible incorrect answer is associated with one or more *bugs* that (likely) exist in a students' knowledge if s/he chose the corresponding incorrect answer. There is an overall mapping from each type of bug evaluated in the assessment to each node in the landscape. For example, take the simple landscape illustrated in Figure 6a. All students start at the node labeled $q0 \in Q$ in the figure. There are one or more bugs, each of which we will refer to as $b_j \in B$, that each node (question q_i) is assessing. Essentially a table with $|Q|$ rows by $|B|$ columns is engineered when the assessment is designed, assigning a Boolean value to each cell in the table indicating which errors are revealed by each question. We use this table to pose two types of questions:

1. *a modeling question*—what is the probability that a student possesses bug b_j, given that they answered question q_i incorrectly, i.e., what is $Pr(b_j|q_i)$?
2. *a prediction question*—what is the probability that a student will answer question q_i incorrectly, given that they possess bug b_j, i.e., what is $Pr(q_i|b_j)$?

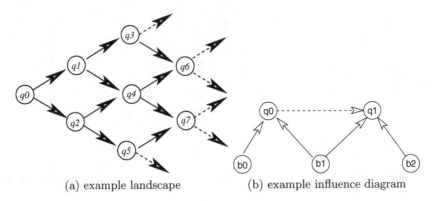

(a) example landscape   (b) example influence diagram

Fig. 6. Agent training structures

Note that there is not a one-to-one correspondence between bugs and questions. Thus we rephrase our two questions:

1. *modeling*—what is the probability that a student possesses the bugs in set B, given that they answered the questions in set Q incorrectly?
2. *prediction*—what is the probability that a student will answer the questions in set Q incorrectly, given that they possess the bugs in set B?

The influence diagram [16,17] shown in figure 6b provides a graphical illustration of this situation. There are two types of variables represented: bugs $\{b_0, b_1, b_2\}$ and questions $\{q_0, q_1\}$. Question q_0 is designed to assess whether a student possesses the bugs in set $B' = \{b_0, b_1\}$; question q_1 is designed to assess whether a student possesses the bugs in set $B'' = \{b_1, b_2\}$.

We use the interaction data set described earlier along with the influence diagram associated with the landscape contained in Figure 1 to compute probability tables, one per cluster. We tally the number of students within the cluster who visited each node and the percentage of them who answered the question incorrectly, indicating particular bugs. Thus, for each cluster, we have a table that indicates how likely it is that a member of that cluster possesses each bug. This is essentially a user model which becomes the heart of the control function for each "cluster agent."

2.4   Evaluating Agents

Finally, we evaluate the efficacy of our methodologies by simulating an assessment using each agent generated. We performed 999 evaluation trials for each agent—since the agents are controlled probabilistically, they will not behave exactly the same way in each simulation run. We evaluate our efforts in four ways. First, we wish to determine how well each agent fits the cluster profile from which it was modeled. The correlation between the behaviors of the trained agents and the groups of humans the agents are emulating is illustrated. Second, we highlight the separation between agents (i.e., distinctiveness of behavior patterns).

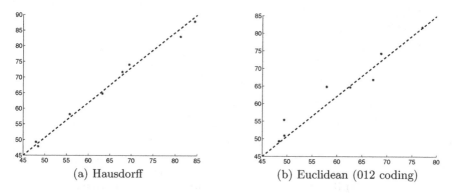

Fig. 7. Correlation between agents and humans. The x-axis represents σ values for the human trajectories within each cluster and the y-axis represents σ values for the trajectories of the corresponding agents.

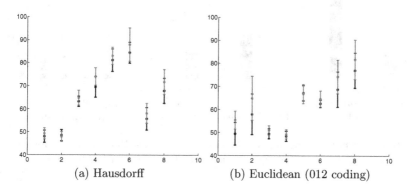

Fig. 8. Separation between agents, using σ

Third, we show that the method of training agents as emulators, rather than taking the raw human data as the basis of user models, provides more coherent results. Fourth, we provide visualizations that compare compounded agent trajectories with humans in corresponding clusters.

Correlation between agents and humans. We use the shape complexity (σ) metric described earlier to compare the relationship between trajectories generated by agents with those generated by humans. Figure 7 plots the average σ for each cluster based on the trajectories exhibited by humans (x-axis) against the corresponding value for trajectories generated by the agent representing each cluster in the 999 evaluation runs. The correlation with the Hausdorff technique is quite high.

Separation between agent behaviors. We also want to compare the separation between agents, recalling our goal to produce a suite of agents, each of which represents different behavior patterns. Figure 8 shows the mean and standard

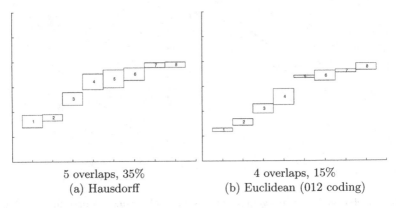

5 overlaps, 35%   4 overlaps, 15%
(a) Hausdorff   (b) Euclidean (012 coding)

Fig. 9. Separation between agents, using y values

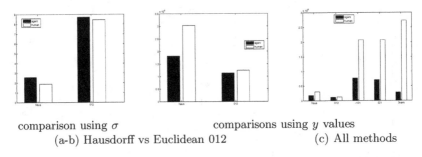

comparison using σ comparisons using y values
(a-b) Hausdorff vs Euclidean 012   (c) All methods

Fig. 10. Coherence with agents, as compared to humans. Plot (a) uses σ for comparison. Plots (b) and (c) use the average variance of y-values for comparison.

deviation of σ for each of the clusters computed for each clustering technique. The black bars represent the clusters based on human trajectories; the grey bars represent the trajectories generated by the agents in the 999 evaluation runs. Once again, the Hausdorff produces superior results because greater distinction between each cluster can be seen in the lefthand plot.

We also examine a different mode of comparison in order to illustrate more clearly the distinction between cluster-based behaviors. Instead of looking at σ, we examine the mean and standard deviation of y values for the trajectories within each cluster. The landscape is shaped such that all trajectories extend horizontally along the same range of x values, but do not cover the same extent vertically. We compare the vertical extent of each cluster by plotting a box outlining one standard deviation around the mean y value. These plots are shown in Figure 8 for both Hausdorff and Euclidean 012 methods. According to this comparison, the Euclidean 012 method appears to perform better than the Hausdorff because it has fewer boxes that overlap and the total amount of overlapping area is less. However, in the Euclidean 012 plot, agent 5's extent is completely subsumed by that of agent 6, which is a worse result.

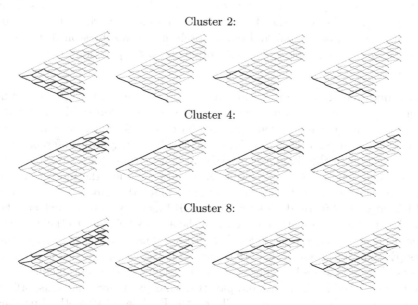

Fig. 11. Comparing agent and human trajectories, clustered using Hausdorff method

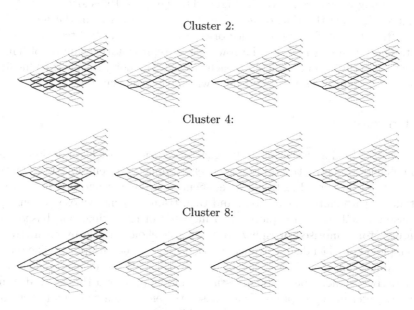

Fig. 12. Comparing agent and human trajectories, clustered using Euclidean 012 method

Another disadvantage of this plotting method is that it does not represent the complete picture because the distribution of y values, weighted by frequency of occurrence, is not normal and more closely matches a Poisson distribution. Thus, we are exploring other methods of comparing separation between agent behaviors.

Coherence within agent behaviors. Figure 10 illustrates the coherence amongst the behaviors of an agent versus humans within a single cluster. Because the agents behave probabilistically, based on the influence diagram and table explained in Section 2.3, a single agent will act (slightly) differently each time it executes in the simulated assessment. The variance of y values is examined for each agent. The smaller the variance of y values, the more coherence there is in the agent's behavior. The lefthand plot (a) compares the Hausdorff and Euclidean 012 methods. The Euclidean 012 method produces better coherence. The plot also compares coherence in the agent's behavior to the coherence across the set of trajectories belonging to the humans in the cluster. In both cases (Euclidean 012 and Hausdorff), there is better coherence in the agent behaviors. This is what we have been striving for. The improvement in coherence is even more marked in the righthand plot (b), which compares the Hausdorff and all four Euclidean coding methods. Clearly the Euclidean 012 method produces superior results.

Visual comparison. Finally, a sampling of trajectories for each cluster and corresponding agent are shown in Figure 11, using the Hausdorff method, and in Figure 12, using the Euclidean 012 method. For each row in the figures, the first (leftmost) plot shows the trajectories (blue lines) over 999 test runs of the agent. The remaining plots in the row show a representative sample of human student trajectories (black lines) for each cluster. The point is that the blue lines represent a composite set of black lines within the same cluster.

3   Conclusion

We have described a methodology for generating agent-based simulations of human behavior in a structured interactive environment. We employed interaction data from an on-line educational assessment environment and created clusters of students with similar behaviors, and then trained agents whose actions typify cluster members. We explored two methods of clustering, one based on a feature-vector comprised of right/wrong answer choices made by each student and employing a Euclidean distance metric to determine groupings. The second is a graphical approach, based on examining the paths students take through the underlying landscape of the assessment and employing a Hausdorff distance metric to determine groupings. From these, we generated a profile for each cluster based on bugs in student knowledge exhibited by the assessment responses. We used these profiles to train agents to emulate cluster members, and finally, we evaluated the efficicacy of these methods by comparing the trajectories produced by agents acting in a simulated assessment to those of humans produced in the real assessment. Our results show that when we use shape complexity (σ) as the basis for comparison, the Hausdorff method is superior to the Euclidean methodologies. Interestingly, attempts to make comparisons based on other metrics, such as the variation in y values aligns the Hausdorff and Euclidean 012 methods more closely. Still, however, the Euclidean 012 method always produces better results than the other Euclidean encodings explored.

Our current work involves extending these methods to other types of data, both from within the education sector and outside it. The type of generic landscape illustrated in Figure 1 can be used to represent the underlying structure of a wide range of interactive environments. Being able to generate agents that emulate human behavior in such environments has broad application and can be used not only for evaluating clustering techniques, as illustrated here, but also for producing controllers for agents that might be deployed as actors within such interactive environments.

Acknowledgments

This work was partially supported by the National Science Foundation under NSF IIP #06-37713 and by the US Department of Education under #ED-07-R-0006.

References

1. VanLehn, K., Ohlsson, S., Nason, R.: Applications of simulated students: An exploration. Journal of Artificial Intelligence in Education 5(2), 135–175 (1996)
2. Sklar, E., Davies, M.: Multiagent simulation of learning environments. In: Fourth International Conference on Autonomous Agents and Multi Agent Systems (AAMAS) (2005)
3. Spoelstra, M., Sklar, E.: Using simulation to model and understand group learning. Agent Based Systems for Human Learning, International Transactions on Systems Science and Applications 4(1) (2008)
4. Sklar, E.: CEL: A Framework for Enabling an Internet Learning Community. PhD thesis, Department of Computer Science, Brandeis University (2000)
5. Cypher, A.: Eager: Programming repetitive tasks by example. In: Proceedings of CHI 1991 (1991)
6. Maes, P.: Agents that reduce work and information overload. Communications of the ACM 37(7), 31–40, 146 (1994)
7. Balabanović, M.: Learning to Surf: Multiagent Systems for Adaptive Web Page Recomendation. PhD thesis, Stanford University (1998)
8. Sklar, E., Blair, A.D., Pollack, J.B.: Training Intelligent Agents Using Human Data Collected on the Internet. In: Agent Engineering, ch. 8, pp. 201–226. World Scientific, Singapore (2001)
9. Hofmann, K.: Subsymbolic user modeling in adaptive hypermedia. In: The 12th International Conference on Artificial Intelligence in Education, Young Researcher Track Proceedings, pp. 63–68 (2005)
10. Mavrikis, M.: Logging, replaying and analysing students' interactions in a web-based ILE to improve student modeling. In: The 12th International Conference on Artificial Intelligence in Education, Young Researcher Track, pp. 101–106 (2005)

11. Merceron, A., Yacef, K.: Educational data mining: a case study. In: The 12th International Conference on Artificial Intelligence in Education (2005)
12. http://www.mathworks.com/products/matlab/
13. Jacobs, N., Blockeel, H.: User modeling with sequential data. In: Proceedings of the 10th International Conference on HCI, pp. 557–561 (2003)
14. Basalto, N., Bellotti, R., De Carlo, F., Facchi, P., Pascazio, S.: Hausdorff clustering of financial time series. Physica A 379, 635–644 (2007)
15. Zhang, Z., Huang, K., Tan, T.: Comparison of similarity measures for trajectory clustering in outdoor surveillance scenes. In: ICPR (3), pp. 1135–1138 (2006)
16. Pearl, J.: Probabilistic reasoning in intelligent systems: Networks of plausible inference. Morgan Kaufmann, San Mateo (1988)
17. Russell, S., Norvig, P.: Artificial intelligence: A modern approach, 2nd edn. Prentice-Hall, Englewood Cliffs (2002)

Evaluation of Automated Guided Vehicle Systems for Container Terminals Using Multi Agent Based Simulation

Lawrence Henesey[1], Paul Davidsson[2], and Jan A. Persson[1]

Department of Systems and Software Engineering, Blekinge Institute of Technology
[1] Box 214, 374 24 Karlshamn, Sweden
[2] Box 520, 372 25 Ronneby, Sweden
{lawrence.henesey,paul.davidsson,jan.persson}@bth.se

Abstract. Due to globalization and the growth of international trade, many container terminals are trying to improve performance in order to keep up with demand. One technology that has been proposed is the use of Automated Guided Vehicles (AGVs) in the handling of containers within terminals. Recently, a new generation of AGVs has been developed which makes use of cassettes that can be detached from the AGV. We have developed an agent-based simulator for evaluating the cassette-based system and comparing it to a more traditional AGV system. In addition, a number of different configurations of container terminal equipment, e.g., number of AGVs and cassettes, have been studied in order to find the most efficient configuration. The simulation results suggest that there are configurations in which the cassette-based system is more cost efficient than a traditional AGV system, as well as confirming that multi agent based simulation is a promising approach to this type of applications.

Keywords: MABS application, automated guided vehicles, container terminal.

1 Introduction

The transport of containers is continuously growing and many container terminals (CTs) are coping with congestion and capacity problems. For instance, the number of Twenty-foot Equivalent Unit containers (TEUs) shipped world-wide has increased from 39 million in 1980 to 356 million in 2004 and growth is projected to continue at an annual rate of 10 per cent till 2020 [1]. Often due to both physical and economic constraints, big container ships are calling on just a few larger ports from which smaller container ships will "feed" containers to other ports in the region. Thus, the number of containers being *transhipped* is increasing.

There is pressure on the management of ports and CTs to find more efficient ways of handling containers and increase CT capacity. Traditional methods for increasing capacity, such as expanding the CT are often not feasible. Thus, other solutions to increase the efficiency and capacity are considered, including the use of automation, e.g., Automatic Stacking Cranes and Automated Guided Vehicles (AGVs) which may also reduce the operational costs for CTs [3].

N. David and J.S. Sichmann (Eds.): MABS 2008, LNAI 5269, pp. 85–96, 2009.

The first AGV system was introduced in 1955 for horizontal transport of materials and AGVs were first used for transporting containers in 1993 at the Delta/Sea-Land terminal located in Rotterdam. There has been much research conducted in various areas incorporating AGVs and CTs (cf. Vis [4], for a comprehensive literature review on AGVs). In two European Union sponsored projects; IPSI (Improved Port Ship Interface) and INTEGRATION (Integration of Sea Land Technologies), a system for handling containers using cassettes and AGVs has been developed but to date has not been used in a CT [5]. The cassettes are steel platforms detachable from the AGV and on which containers can be set. A possible advantage of using cassettes is their ability to act as buffer, since containers can be placed on it without an AGV being present. Figure 1 shows an IPSI AGV and illustrates the different parts of a CT.

 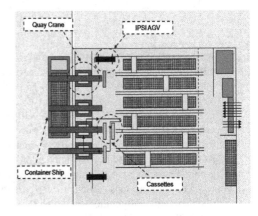

Fig. 1. An IPSI AGV transporting a cassette loaded with two containers (left) and a sketch of a container terminal (right)

To evaluate this new development in container handling, we will compare the IPSI AGV system to a "traditional" AGV system, which will be referred to as T-AGV. We perform a comparative analysis of the transport of containers between ship-side operations to the operations in the stacks located in the yard . We also study the effect of the number of cassettes allocated to the IPSI AGV. The developed model can also assist in analyzing aspects, such as, crane assignment to ship and AGV assignment to cranes.

Because of complexity and capital and construction costs, simulation models have been used for understanding behaviour and testing strategies in CTs, e.g., see [6], [7] and [3]. We have developed a multi-agent based simulator (MABS) for comparing the performance of the two AGV systems according to a number of criteria, e.g., service time for a ship, utilization rate for the CT equipment, and operating cost. The choice of using MABS is based on the versatility in simulating complex systems and perceived simplicity in the structure preserving modelling of the different entities in a CT. Parunak et al. [2] compared macro simulation and micro simulation approaches and pointed out their relative strengths and weaknesses. They concluded that "...agent-based modeling is most appropriate for domains characterized by a high degree of localization and distribution and dominated by discrete decision. Equation-based modeling is most naturally applied to systems that can be modeled centrally, and in which the dynamics are dominated by physical laws rather than information processing." As a CT has a high degree of localization and distribution and is dominated by discrete decision, we found agent-based modeling a promising approach worthy to investigate.

The remainder of the paper is organized as follows; in section 2 a description of the problem is provided. In section 3, the methodology and model is presented. Section 4 provides a description of the simulation experiments. The results are presented and discussed in section 5. Conclusions and future work are provided in section 6.

2 Problem Description and Model Assumptions

A CT is a place where ships will be berthed so that containers can be unloaded and/or loaded by Quay Cranes (QCs). The CT are often viewed as an interface for transport of containers between modes of transport linking the landside with the marine-side [9]. In addition, CTs can serve as temporary storage in the transhipping of containers from one ship to another ship. Ship owners often demand a fast turn-around time since more time for sailing imply more opportunities for extra voyages and more revenue. Further, from a logistics perspective, improving the transport of containers within the CT may help in decreasing the total transport time and cost of transporting cargo in containers. CT managers seek to meet requirements of fast turn-around time and try to optimise the use of their terminal resources, such as, quay cranes for which the capital costs can be € 7 million or more [10], and transporters, such as AGVs. Many CT managers view the interface between the QCs and the yard as the most critical planning problem [11].

In the scenario studied, a ship arrives at a CT with a number of containers to be unloaded and another set of containers are loaded onto the ship before it departs. The unloaded containers are to be transported from the QC area to stacks in the yard and the containers to be loaded are picked-up from stacks and transported to the QC area. We will refer to the QC areas where containers are stored temporarily as the Buffers. The transportation between the buffers and the stacks is carried out by AGVs (either T-AGVs or IPSI AGVs with their accompanying cassettes) and we call the time it takes to perform it (including the return without container(s) as well as the load and unloading operations) the AGV cycle time. This definition is similar to the one used in a study comparing a Straddle Carrier system with an Automated Stacking Crane system by Vis [12]. The stacks are located in different areas of the yard and therefore have varying distances to the QC, implying different transport times. We model this by letting the AGV cycle time have random component for each transport. We also consider the time for the unloading and loading of containers from and to a ship by a QC, called the container handling time.

The technical specifications of the two AGV systems considered are presented in Table 1. The IPSI AGV is slightly faster than a T-AGV and has a higher loading capacity as containers cannot be stacked on the T-AGV. The lifting time for an IPSI AGV (the time for it to move under a cassette and lift it off the ground for transporting) is approximately 15 seconds. The T-AGVs do not have a corresponding lifting time since containers are directly loaded on top of them. The initial purchasing cost is provided by industrial partners and serves as an estimate.

Table 1. Specifications of the AGV systems (kr = Swedish kronor)

	IPSI AGV	T-AGV
Speed (both empty and loaded)	20 km/h	15 km/h
Capacity	82,000 kg	55,000 kg
Maximum container capacity that can be transported	4 TEU (either 4 x 20' or 2 x 40 containers)	1 TEU (one 20' container) or 2 TEU (one 40' container)
Initial purchasing cost	4,5 million kr	2,7 million kr
Additional costs (one cassette)	8,000 kr	Not applicable

3 Methodology and Simulation Model

In our model there are entities which have a number of attributes and operations associated to them and some of which can communicate with each other. The entities of the real world that we model are: ships, QCs, their buffers, AGVs (IPSI AGV or T-AGV), cassettes and containers. These entities (in some cases including the persons operating them) have to coordinate with each other to complete the main task, which is the unloading/loading a ship. Therefore we model the AGVs, the buffers and the QCs as agents, which. during the simulation perform their task in parallel, replicating the activities in real world CTs. The simulation model was implemented using DESMO-J, an open source library for the JAVA programming language [14]. DESMO-J provides a runtime process based simulation engine that can be used to map port entities to software entities and to simulate the coordination of these process.

The entities use the Contract Net protocol to coordinate tasks. This protocol is used because of its ability to distribute tasks and self-organise a group of agents [15]. The protocol is suitable since our model describes tasks that can be characterised as hierarchical in nature and are well-defined. The Contract Net protocol implies that one agent will take the role of a "manager", which initiates a job to be performed by one or more other agents. The job may require that a number of participating agents respond with a proposal and the manager will accept a proposal and confirm it to a selected agent and reject the other proposals. This protocol seems to closely reflect the operational decisions that are made by the actual workers in the CT, especially when a foreman will communicate via radio with drivers and QC operators. In our case, the buffer agents act as "managers" of the Contract Net protocol and the AGV agents are bidding and performing the jobs.

The system that we have modelled is illustrated in Figure 1 (right-hand side). We have focused on modelling the operations that involves the QCs and the AGVs that transfer containers between the quay and stacks. The main difference between the IPSI AGV and the T-AGV is the absence of cassettes. The containers need to be placed directly on the AGV in the T-AGV system. A QC will wait for an available T-AGV to place a container on it rather than placing the container on the ground and then having to later pick it up and place it on a T-AGV, in order to reduce extra moves by the QC. We have followed a general simulation process as described by Law and Kelton [8] and therefore we are testing a prototype with real data.

The entities that are modelled are:

- *Ship:* contains the containers.
- *Quay cranes:* used to unload and load the containers.
- *AGVs* (IPSI AGVs or T-AGVs): used to transport container, from/to QC from/to a container stack.
- *Buffer:* Pick-up and drop-off area behind the QC that temporarily stores containers on cassettes or AGVs.
- *Cassettes:* A set of cassettes bound to a QC.
- *Containers*

As mentioned earlier, the QCs, Buffers and the AGVs are modelled as agents, whereas the ships, cassettes and containers are modelled as objects.

3.1 Quay Crane (QC) Agent

The QC is responsible for the unloading and loading the containers from and to the ship. The number of QCs serving a ship is specified by the user. Each QC has the following attributes:

- a set of AGVs (IPSI AGVs or T-AGVs) assigned to it
- a set of cassettes (for the IPSI AGVs)
- a buffer area where its cassettes or T-AGVs can be placed for loading/unloading containers
- container handling time for (un)loading a container. In a real CT the container handling time will vary for each container, which we simulate by a computer-generated random number using a linear congruence method chosen in a range specified by the user.

Moreover, the following are recorded for each QC:

- the number of containers unloaded from the ship and loaded to the ship
- the time it has been working (not including the idle time).

A QC performs the following functions

- When unloading, the QC will unload a container from the ship if there is a cassette (or T-AGV) with free space in the buffer area. If not, it waits until there is one available.
- When there are containers to be loaded available in the buffer area, the QC will load one container at a time to the ship.
- The QC finishes working when there are no containers to be unloaded or loaded for the ship.

3.2 Buffer Agent

A buffer is assigned to a specific QC and the buffer is responsible for allocating free AGVs to either pick-up a container and move it or move empty (to pick-up container(s) at another location). The buffer is also responsible for the QC to stop unloading if there is no cassette available or the cassette is full and to stop loading if there is no containers available on cassettes or on AGVs. The Buffer agent will communicate with the AGVs and assign an AGV that is free to pick up a cassette (for the IPSI-AGV). Once a cassette is available the buffer agent will ask the QC to start working. When using T-AGVs, the buffer agent tries to find a free T-AGV and assign a container to that T-AGV. If no T-AGV is free then it waits until a T-AGV is free at the buffer. Pseudo-code describing the unloading and dispatching strategy of the buffer agent for the cassette-based system is given below:

```
WHILE still containers to unload DO
   IF cassette available that has room for more containers  THEN
      Ask QC to unload a container and place it on cassette
      Ask all AGVs for their status
      Wait for status reports
      IF AGV idle THEN
         Ask that AGV to fetch the loaded cassette
   ELSEIF cassette is full THEN
      Ask QC to stop unloading
      REPEAT
         Ask all AGVs for their status
         Wait for status reports
      UNTIL at least one AGV is idle
      Ask the idle AGV to fetch the loaded cassette
ENDWHILE
```

3.3 AGV Agent (IPSI AGV and T-AGV)

Each QC has a number of AGVs assigned to it. This value is specified by the user before the start of the experiment. An AGV is responsible to transport containers between a QC Buffer and container stacks. Each AGV has the following attributes:

- a state ("free" or "busy")
- a cycle time for transporting a cassette/container from the buffer to the stack and return back to the buffer. Or vice versa during the loading phase. In a real CT the AGV will have varying transport times, which we simulate by a computer generated random number using a uniform method chosen in a range specified by the user.

Moreover, the following are recorded for each AGV:

- the number of containers transported to/from the stack
- the time it has been working (not including the idle time).

4 Experiment Description

The input parameters are stored in a text file from which the simulator reads the parameters. The output of the simulation is a set of files which contains information of all events taken place during the simulation. A trace file contains the overall performance of each QC, AGV and cassette involved in the simulation. The performance criteria that are used for evaluating and comparing the CT transport systems are:

- *Service Time*: is the time it takes to complete the unload/load operations for a ship, also known in the maritime industry as "turn-around time".
- *Utilization Rate*: *Active time / Service Time (Active time + Idle time)*. *Active time* is the time a piece of CT equipment is busy, e.g. moving a container from the QC to a stack, and *Idle time* is the time that it is not working. The utilization rate for the following equipment is recorded: *QC, AGV and Cassette*.
- *Throughput*: Avg. number of containers handled per hour during Service time for: *QC, AGV and Cassette*
- *Total Cost*: Equipment cost for serving a ship is calculated in the following ways (OPEX = operating cost per hour for a unit of CT equipment):
  - *QC:* number of QCs x OPEX for QC x Service Time.
  - *AGV:* number of AGVs x OPEX for AGV x Service Time.
  - *Cassette:* number of cassettes x OPEX for Cassette x Service Time.
  - *Total Cost: QC costs +AGV costs + Cassette costs*

4.1 Scenario Settings

The scenario settings were based upon data provided by industrial partners. The results from the simulations are average values. Thus, a sufficient number of runs are needed in order to get a valid estimation. The cycle times used in the simulation have been determined from prior analysis in which the stack distances and maximum speeds of the AGVs were tested. We used an approximation method to calculate the minimum number of simulation runs required in order to obtain results from a simulator with small enough statistical errors. The approximation method [16] has been applied by Vis et al. [17] in vehicle allocation at a container terminal. In this method, data from a limited number of replications is used to approximate the required minimum number of replications in the actual experiment (denoted i) such that the relative error is smaller than γ ($0 < \gamma < 1$) with a probability of 1-α. The i value can be calculated from:

$$i \geq S^2(i)\left[z_{1-\alpha/2} / \gamma' \overline{X}(i)\right]^2 \tag{1}$$

where $S^2(i)$ is the variance of the trial sample, $z_{1-\alpha/2}$ is the 1-α/2 percentile of the normal distribution, $\overline{X}(i)$ is the trial sample mean value, and $\gamma' = \gamma/(1+\gamma)$. Based on our trial sample we found that to obtain the error smaller than 2% with a probability of 95% that the number of generated replications would be sufficient at 100 for all experiments conducted in this paper. In the simulation experiments, we use the settings listed in Table 2 for serving a single ship.

An "average ship" is used in which 493 containers are to be either unloaded or loaded, and there are 3 QCs to serve the ship. Each QC is assigned 1 to 5 AGVs and has a container handling time randomly generated for each container ranging between 1 to 2 minutes. The numbers of cassettes are from one to four per IPSI AGV. Each IPSI AGV has a travel cycle time that is randomly generated for each cassette transported ranging between 3 to 6 minutes. The T-AGVs posses a random travel cycle time ranging between 3 to 5 minutes. Cycle time for the IPSI AGV includes the lifting of a cassette, transport it from a QC to a stack, detach the cassette and then return to the QC with an empty cassette; or the cycle time for the opposite direction, i.e., transporting from the stack to the QC. Cycle time for T-AGVs is similar to IPSI AGVs but does not have a lifting time or transport a cassette.

Table 2. Settings experimented in the simulator for a single ship

Input Settings for Scenario	AGV Type	
	IPSI AGV	T-AGV
No. Containers	493	493
No. QCs	3	3
No. Cassettes assigned to an QC for each IPSI AGV	1- 4	n.a.
No. AGVs per QC	1-5	1-5
Container handling time for QC	1-2 min	1-2 min
Travel cycle time for AGV	3-6 min	3-5 min

5 Simulation Results

Simulation experiments were conducted to evaluate various combinations of allocated terminal resources three QCs. Ship service time results are presented in Figure 2, for different number of AGVs and cassettes used (the exact values are given in Table 3). They suggest that ship service time is generally faster for IPSI AGVs than for T-AGVs. When three or more IPSI AGVs with two or more cassettes each, the service

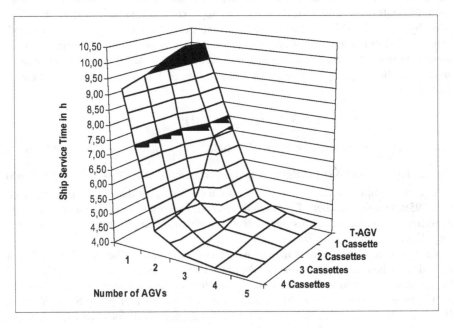

Fig. 2. Simulation results for ship service time

time is close to its smallest value and instead the capacity of the QCs becomes the bottleneck. Ship service time results for the T-AGV system are similar to the IPSI AGVs when assigned with one cassette. The ship service time appears to be faster after two IPSI AGVs are assigned with two or more cassettes, average ship service time is 5,13 hours. The fastest ship service time is 4,10 hours when using four cassettes and either four or five IPSI AGVs. The use of an additional IPSI AGV when using four cassettes appears not to influence the ship service time.

Table 3. Average ship service times (*and standard deviation*)

No.AGV	T-AGV	1 Cassette	2 Cassettes	3 Cassettes	4 Cassettes
1	10,03 *(0,14)*	10,08 *(0,12)*	9,85 *(0,13)*	9,45 *(0,14)*	9,23 *(0,13)*
2	7,15 *(0,05)*	7,03 *(0,06)*	5,13 *(0,06)*	4,90 *(0,07)*	4,78 *(0,06)*
3	4,65 *(0,06)*	4,58 *(0,07)*	4,32 *(0,05)*	4,23 *(0,06)*	4,23 *(0,06)*
4	4,33 *(0,06)*	4,30 *(0,06)*	4,27 *(0,07)*	4,15 *(0,06)*	4,10 *(0,06)*
5	4,20 *(0,06)*	4,20 *(0,06)*	4,18 *(0,06)*	4,13 *(0,07)*	4,10 *(0,06)*

From the simulation experiments, we can compare the QC utilization rates in Table 4. Generally, we see that the more transport equipment is available, the higher is the QCs' rate of utilization. The rate of QC utilization becomes close to one when using three or more IPSI AGVs with at least two cassettes per AGV.

Table 4. Comparison of QC Utilization Rates

No. AGVs	T-AGV	1 Cassette	2 Cassettes	3 Cassettes	4 Cassettes
1	0,41	0,41	0,42	0,43	0,44
2	0,57	0,58	0,80	0,84	0,86
3	0,88	0,89	0,95	0,97	0,97
4	0,95	0,95	0,96	0,99	1,00
5	0,98	0,98	0,98	0,99	1,00

AGV utilization rates are presented in Table 5. We see that the utilization rate is close to 1 when only one AGV is used, that is, the QC is able to keep the AGV busy. When more AGVs are added, the utilization rate decrease and the AGVs spend more time being idle. In comparing T-AGVs with the IPSI AGVs, there is a recorded higher level of utilization when IPSI AGVs each have two or more cassettes. Utilization rates for the IPSI AGVs decrease in smaller increments as the number of cassettes increase.

Table 5. Comparison of AGV Utilization Rate (IPSI AGV and T-AGV)

No.AGVs	T-AGV	1 Cassette	2 Cassettes	3 Cassettes	4 Cassettes
1	0,957	0,962	0,995	0,997	0,991
2	0,687	0,673	0,939	0,971	0,983
3	0,659	0,680	0,730	0,755	0,744
4	0,543	0,559	0,570	0,575	0,578
5	0,477	0,484	0,476	0,469	0,468

An increase in the number of cassettes and AGVs adds extra capacity for transporting containers. The extra capacity provided by cassettes may be viewed as a 'floating buffer', which allows the IPSI AGVs to decouple the load of containers on a cassette and fetch another cassette. This activity assists in lessening the idle time of the QCs so that they can be more productive. Thus, from the above results one can conclude that it is useful to introduce a certain amount of IPSI AGV and cassettes in the simulation to make the crane busy throughout the simulation. As crane operating cost is higher than the AGV operating cost, these results can be helpful for CT management in deciding, e.g., how many cranes, IPSI AGVs and cassettes to be allocated to a ship. We shall now compare operating costs for the different configurations.

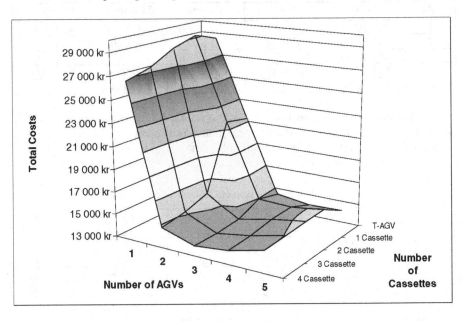

Fig. 3. Total operating costs for serving a ship

Table 6. Total operating costs for serving a ship

No. of AGVs	T-AGV	1 Cassette	2 Cassettes	3 Cassettes	4 Cassettes
1	28 535 kr	29 206 kr	28 545 kr	27 400 kr	26 786 kr
2	21 257 kr	21 649 kr	15 816 kr	15 112 kr	14 766 kr
3	14 424 kr	14 939 kr	14 090 kr	13 837 kr	13 856 kr
4	14 001 kr	14 796 kr	14 707 kr	14 330 kr	14 182 kr
5	14 112 kr	15 215 kr	15 186 kr	15 035 kr	14 945 kr

In Table 6 and Figure 3 the total operating costs for employing the CT equipment is presented. In determining the total operating costs the hourly operating cost is multiplied by the number of CT equipment type employed, which is then multiplied by ship service time. The assumed hourly operating costs (including depreciation, maintenance, labour and fuel) used in the calculations are (in Swedish kronor per hour):

- QC: 905 kr/hour
- T-AGV: 43 kr/hour
- IPSI-AGV: 60 kr/hour
- Cassette: 0,50 kr/hour

In comparing the total operating costs, the addition of more AGVs and cassettes leads to lower costs up until three IPSI AGVs and three cassettes are employed. The total costs when adding further equipment increase, but the time gained do not compensate for the extra cost. Thus, from a cost perspective the best choice in the scenario studied would be to use three IPSI AGVs with three cassettes each. From a ship service time perspective, on the other hand, an additional AGVs and an additional cassette for each AGV would give even better results (but inly slightly). Thus, a CT manager would have to make the final decision of how to balance the quality of service provided with the total costs.

6 Discussion, Conclusion and Future Work

The simulation experiments have provided much insight into the properties of various terminal equipment types that can be used in container terminal operations. The cassette-based system posses some advantages in that it can act as a 'floating' buffer, meaning that it can allow the QCs to keep unloading/loading and not having to wait for an AGV to be available. Waiting time is lower for the QCs and thus they are obtaining better utilization rates. The initial results from the prototype AGV simulator provide some interesting observations useful for determining the number of machinery units to allocate for serving a ship. The simulation experiments that we have conducted are also creating further questions that require more investigation. Naturally there is a trade-off to be expected between service time and the costs for purchasing and operating equipment.

Compared to traditional simulation approaches, the MABS approach can provide finer granularity in modelling the entities and having them communicate and coordinate amongst other entities. In particular, we found this useful in the studied application where the characteristics of different types equipment need to be captured, and where equipment is distributed in physical space and the need to coordinate their activities is essential. More generally, we argue that MABS seems a promising technology for evaluation and comparison of different automation approaches. Elder [13] mentions some general advantages of simulation methods over applying queuing theory. In addition, we found it difficult to use queuing theory to model activities such as the AGV dispatching strategies (in this study carried out by the Buffer agent) that are used when applying cassette-based systems.

For future work we would study other models for container handling time by the QC and the cycle time for AGVs (e.g., including stoppages caused by malfunctioning equipment, etc. which affects the productivity at a real CT). Another topic worth studying is different dispatching strategies for allocating containers to AGVs and cassettes. We plan to extend the model in several directions, e.g., including the unloading/loading taking place at the stacks, more detailed modelling of the AGV movements, etc.

Acknowledgements

This work has been partially funded by Karlshamn Municipality. The following port industry representatives have provided useful information necessary for the development of IPSI AGV Simulator; TTS AB in Göteborg, Sweden, Lennart Svensson, Bjørn O. Hansen and Michel Lyrstrand. Much appreciation is given to Khurum Aslam for related work with the DESMO-J simulator and to Dr. Piotr Tomaszewski.

References

1. Davidson, N.: A global capacity assessment and needs analysis. In: 39th Terminal Operating Conference. Informa Plc., Antwerp (2005)
2. Parunak, H.V.D., Savit, R., Riolo, R.L.: Agent-Based Modeling vs. Equation-Based Modeling: A Case Study and Users' Guide. In: Sichman, J.S., Conte, R., Gilbert, N. (eds.) MABS 1998. LNCS, vol. 1534, pp. 10–26. Springer, Heidelberg (1998)
3. Ioannou, P.A., Kosmatopoulos, E.B., Jula, H., Collinge, A., Liu, C.-I., Asef-Vaziri, A.: Cargo Handling Technologies. Technical Report. Department of Electrical Engineering, University of Southern California: Los Angeles, pp. 1–147 (2002)
4. Vis, I.F.A.: Survey of research in the design and control of automated guided vehicle systems. European Journal of Operational Research 170(3), 677–709 (2006)
5. TTS AB Port Equipment, http://www.tts-marine.com
6. Bruzzone, A.G., Giribone, P., Revetria, R.: Operative requirements and advances for the new generation simulators in mulitmodal container terminals. In: Proceedings of the 1999 Winter Simulation Conference. Society for Computer Simulation International (1999)
7. Hayuth, Y., Pollatschek, M.A., Roll, Y.: Building a Port Simulator. Simulation 63(3), 179–189 (1994)
8. Law, A.M., Kelton, W.D.: Simulation Modeling and Analysis, 3rd edn. McGraw-Hill International, Boston (2000)
9. Rida, M., Boulmakoul, A., Laurini, R.: Calibration and validation of container terminal simulation. In: Proceedings of the 3rd IEEE International Symposium on Signal Processing and Information Technology, pp. 774–777 (2003)
10. De Monie, G.: Environmental Scanning in Ports. In: ITMMA Private Public Partnerships in Ports, Antwerp, Belgium (2005)
11. Henesey, L., Davidsson, P., Persson, J.A.: Agent Based Simulation Architecture for Evaluating Operational Policies in Transshipping Containers. In: Fischer, K., Timm, I.J., André, E., Zhong, N. (eds.) MATES 2006. LNCS (LNAI), vol. 4196, pp. 73–85. Springer, Heidelberg (2006)
12. Vis, I.F.A.: A comparative analysis of storage and retrieval equipment at a container terminal. International Journal of Production Economics 103(2), 680–693 (2006)
13. Elder, M.: Simulation beats Queuing, SIMUL8 (2005)
14. DESMO-J, http://www.desmoj.de/
15. Clearwater, S.H.: Market-Based Control: A Paradigm for Distributed Resource Allocation. World Scientific, Singapore (1996)
16. Law, A.M., Kelton, W.D.: Simulation Modeling and Analysis. McGraw-Hill International, Boston (2000)
17. Vis, I.F.A., de Koster, R., Savelsbergh, M.W.P.: Minimum vehicle fleet size under time-window constraints at a container terminal. Transportation Science 39(2), 249–260 (2005)

MASFMMS: Multi Agent Systems Framework for Malware Modeling and Simulation

Rohan Monga and Kamalakar Karlapalem

International Institute of Information Technology

Abstract. The Internet and local area networks, which connect many personal computers, are also facilitating the proliferation of malicious programs. Modern malware takes advantage of network services like e-mail and file sharing to proliferate. Existing simulation environments use biological models or their variants for explaining the patterns of proliferation of malicious programs. This paper aims to provide a framework that enables the modeling of security threats using multi agent systems. Multi Agent Systems Framework for Malware Modeling and Simulation (MASFMMS) provides a generic environment for modeling security weaknesses and their exploitation in a computer network. We present various scenarios of exploits that are prevalent in real life and show how they can be simulated in MASFMMS.

1 Introduction

Most organizations today have computer networks to cater to their business needs. These networks are valuable assets and a lot of money is spent on their security. The convenience of connectivity across computers has its hidden dangers. A wide variety of malicious programs threaten the security and integrity of data data resides on these computer networks[12]. In this paper, the term *"security weakness"* means vulnerability present in a computer due to which proactive attacks on the computer have a potential for success. Security weaknesses include buggy software, lack of firewalls or antivirus programs, or lack of care taken by the owner of the computer to safeguard it. A quick scan of documented attacks [1] reveals that these are the causes of almost all security exploits.

1.1 Motivation

A wide variety of malicious programs like viruses, worms, Trojan horses etc. have been written over the years. These programs are unique in terms of the way they spread, the kind of damage they intend and the way they locate new targets. According to CERT, over 36,000 security vulnerabilities have been documented so far. There are over 36 different types of malicious programs and over 7200 different malicious programs [1] and these lists are growing everyday.

Modeling the exploitation of a security weakness is a challenging task due to a large number of factors involved in the process. Weaver et al. have defined a taxonomy and described in detail the parameters that are required to adequately

N. David and J.S. Sichmann (Eds.): MABS 2008, LNAI 5269, pp. 97–109, 2009.
© Springer-Verlag Berlin Heidelberg 2009

model a worm[13]. Most exploits are due to flawed software, but that is not the sole reason for infection. Delay in the installation of patches and naively opening infected email attachments are some of the other reasons for infection. Malware, in fact, is difficult to model. Malicious programs use complicated techniques to find new targets, infect them, stay undetected and then infect other computers. There is a need for many different models to explain the behavior and to reason about the spread of a new security exploit.

Due to the sheer number of security weaknesses and their wide variety, modeling and predicting the behavior of malware has been difficult. Older models like the classical worm propagation model [3], Kermack-McKendrick Model [4], two-factor worm propagation model [5] try to apply biological propagation models and their variations to explain the propagation of worms. These models miss out on some of the parameters described by [13] and hence would fall short in their results. For example, they cannot be applied to viruses, Trojans or other malware that do not intend to spread rapidly. The user behavior and target location algorithms (of viruses and worms) cannot be simulated. There is a need to experiment with new models and build fine grained simulators to test their accuracy. Hence we need a framework which can adequately and realistically allow the development of a model to explain the exploitation of a particular kind of security weakness. This framework should be able to provide comparative results between two models and also be able to provide a platform for studying the effects of multiple malware infection simultaneously.

Simulations like these can give information about the rate of proliferation of malicious programs; they will give system administrators a prior knowledge of vulnerabilities of their networks and help them determine the types of malicious programs that are most likely to infect their networks. This framework can also help security experts simulate new malicious programs and validate their results against real scenarios.

1.2   The Need for Multi Agent Systems

The current models are inadequate because they do not treat the computer as an autonomous entity. They model the effect of malicious programs by continuous differential equations. Multi Agent Systems (MAS) give us the advantages of decentralized data and asynchronous computation. The freedom of decision making lies with the agents. In computer networks, the computers are autonomous, like the agents in MAS. The computers communicate and coordinate with each other to accomplish tasks given by users much like in MAS. Similarly, in MAS, the environment provides a channel to communicate but places restrictions on actions that the agents can perform. The computers can communicate with each other as long as they follow the network protocol.

These similarities make MAS well suited for a framework to model a computer network environment. The models developed in such a framework will be simple to understand and give results which are close to reality.

1.3   A Computer as an Agent

A computer on a network has many aspects which need to be represented to make the simulation realistic and meaningful. Primarily, we need to model (1) the software, (2) the security software installed on the computer, and (3) the user. The software installed on the computer gives us information about various flaws and the exploitable logic causing the vulnerability that can be used to infect the computer. Local security software, for example, firewalls or anti-viruses can prevent infection even when buggy software is present. Hence if we do not model the security software separately, our simulation results will not be realistic. The user of the computer is the most important aspect in translating a computer on the network to an agent in the simulation. The user decides the uptime of the computer, the runtime of different programs, and the installation of patches. MASFMMS provides methods to define these parameters (explained in Sect. 3.2).

The rest of the paper is organized as follows. In section 2, we describe how agents and their environment are defined in MASFMMS. We also visit the process used by malicious programs for infection. In section 3, we dive into the architectural details of MASFMMS. In section 4, we describe how models are defined in the framework and present some experimental results.

2   MASFMMS Overview

2.1   Computers

MASFMMS models computers as agents. In the paper, "agent" and "computer" are used interchangeably. The computer network is modeled as the environment. The agents are autonomous entities, and there is no central guidance or command system. They can coordinate with each other by sending messages via the environment.

The agents are uniquely identified by an IP address, just like computers on the computer network. MASFMMS allows for the specification of the following attributes for all the computers: (1) the list of software installed on the computer along with their usage statistics, (2) the list of security software, and (3) the probabilities of the user applying a patch or removing the malware.

Once all the agents are initialized, MASFMMS starts simulating the normal functioning of the computer. It simulates the execution of some software depending on the usage statistics specified. Some of this software is vulnerable and can be infected by the malware which is being studied. Upon infection, the agent is controlled by the malware. The malware makes the agent malfunction and can even take it off the network. If the infected agent has not been taken of the network, it starts to look for possible targets by scanning other agents in the environment and figuring out what software is installed on them. In the real world, this is equivalent to malicious programs running port scans, sniffing or exploiting network protocols to generate errors [6]. The framework allows this by exchange of lists of software amongst agents via the messaging system.

Security software like firewalls and antivirus programs do not allow free exchange of such information. Firewalls can be configured to drop packets and not send information which can lead to infection. Similarly, MASFMMS allows the security software installed on the agents to mask certain information. For example, it is possible to find the kernel version of the OS by TCP/IP fingerprinting [7], and if some firewall is not allowing the computer to send any ICMP error messages, the malware would be unable to figure out this information. The security measures modeled in MASFMMS would restrict information based on their configuration. (See Sect. 2.4).

The user can install patches or pro-actively remove the malware. The framework allows the modeling of the basic features of user behavior. The patching rate and the general death rate of the malware can be specified with probabilistic distributions. However, these are system wide properties and specifying them at the agent level is not supported. The parameters for user behavior are simulation wide but can easily be extended to be different for each agent. As an experiment, we programmed P. Steel's model of user procrastination (See 4.5)

2.2   Network

MASFMMS assumes that all the computers in the computer network are free to interact directly with each other. We consider scenarios where the network allows direct connections, sniffing and network stack analysis, ICMP or other similar error messages or at least one of these. This means that either the computers lie in the same subnet or if they lie in different subnets they can send data to each other without screening. If the computers are unable to communicate with each other directly, let's say due to the presence of a proxy, they can be considered to be in separate networks. This assumption is valid because malicious programs in one part of the network will not be able to infer information as described in Sect. 2.1 and hence would not be able to infect the other part of the network.

MASFMMS does not allow for a micro level simulation of the computer network. This means that core network related parameters like topology, packet level analysis, network speed etc. cannot be modeled. The network itself is considered to be a black box in terms of applications of one computer communicating with applications on another computer via standard network protocols. The emphasis here is on providing a framework for modeling different kinds of malware, their spread patterns and allowing simulation of various parameters which might affect these. Scenarios that deal with physical network simulations are beyond the scope of this paper.

2.3   Security Weaknesses and their Exploits

To model varied kinds of security weaknesses and their innovative exploitation techniques, MASFMMS allows malware logic to be supplied to it externally. Once all the agents have been initialized and the simulation is running, the malware can be injected in the system to study its effect. MASFMMS allows the malware to be specified along the following parameters: (1) Malware name, (2) Threat level, (3) Patching information, (4) Target location logic, (5) Infection logic, (6) Post-infection logic, and (7) List of vulnerable software. Threat level is a measure of how well current antivirus technology can keep the malware from spreading. Generally, older malware techniques are well contained; new threat types or complex malware can be more difficult to contain and thereby qualifies as a greater threat. Threat levels are defined as Category I (threat is well-contained), Category II (threat is partially contained), and Category III (threat is currently uncontainable). Software that the malware can exploit is specified in the list of vulnerable software and patching information specifies how to fix those flaws to prevent further infection. Target location logic and infection logic specify factors which the malware considers for the location of a possible host and how it plans to infect it. Infection techniques would typically include exploiting flaws in the software on the computer. Post-infection logic is what the malware does upon infection. It could look for new targets, remove software from the computer or bring down the computer all together. For example, Code Red v2 [2] generated 100 threads. Each of the first 99 threads randomly chose one IP address and tried to set up connection on port 80 with the target machine: this is the target selection algorithm. If the connection was successful, the worm would send a copy of itself to the victim web server to compromise it and continue to find another web server: this is the infection algorithm. If the victim was not a web server or the connection could not be setup, the worm thread would randomly generate another IP address to probe.

2.4   Security Software and the Infection Process

MASFMMS helps to simulate the process of infection, i.e., transfer of the malicious program from the infected host to a vulnerable one. Let's take the example of Code Red v2 [2]. Code Red worm exploited a Windows IIS vulnerability for infection. It had a random number generator for target location and was extremely virulent. It infected millions of computers and caused massive damage. Code Red v2 has been used time and again for worm propagation studies[5][9]. For simplicity, we consider only two computers A and B in the network and consider a couple of test cases comparing how infection would spread in reality vs. MASFMMS.

In the first case, we assume that neither of the computers have any security software installed and A is assumed to be seeded with infection at the start of the simulation. Given this situation, in reality, the worm first finds out about the existence of B. It could do that by using a variety of techniques, such as sniffing the network, randomly checking different IP addresses etc. Code Red v2 randomly generates targets and lets say B gets chosen. It then confirms if it can be infected by concluding the existence of the vulnerable version of Microsoft IIS. After checking, it exploits the vulnerability and transfers its code to B and continues. MASFMMS allows a similar infection methodology. The worm logic in agent A randomly generates addresses and lets say the address of B is generated. Then A asks B for a list of all the software installed on it. When B sends back the list and Microsoft IIS is present on it, the worm in A realizes that B can be infected. It then sends a copy of its own infection logic, target location logic and post-infection logic to agent A. On reception of these parameters, B is infected because of the lack of security software and the vulnerability. B starts infecting other computers.

In the second case, let's assume B has an antivirus or a firewall that protects it from Code Red v2; or the Microsoft IIS server was patched and no longer vulnerable. In this scenario, Code Red v2 in A is unable to locate the security weaknesses in B and is not be able to infect B, or it tries to infect and fails. MASFMMS allows a similar process. When B is requested for its list of software, after a process of discovery and transfer of malicious code as explained above, it consults its security logic. This security logic is described by the security software installed on the agent. The security logic has a list of known vulnerabilities that it protects the agent from. The security logic consults this list and finds Microsoft IIS. It then masks the information being sent to A such that Microsoft IIS is not there on the list. In another scenario, B gets infected by sending the complete list, but the security logic in B finds that the incoming information is a known exploit and rejects it. These scenarios are similar to the real scenarios in which a firewall drops packets or patches prevent infection by removing the vulnerability. Hence either A is unable to infer that B is vulnerable or it tries to infect and B rejects the code due to its security logic thereby staying safe.

These are simplistic scenarios; complex simulations, however, consist of thousands of agents. The malware also tends to be more complicated; examples of such simulations are described in the experiments section.

3   MASFMMS

3.1   Architecture of MASFMMS

The architecture of MASFMMS is shown in Fig. 1. It consists of the following: (1) Database, (2) Agent pool, (3) Thread pool, (4) Scheduler, (5) Message queue, and (6) Malware cache.

Database stores initialization information about all agents. It also stores descriptions of malware. The parameters along which malware is described are explained in Sect. 2.3.

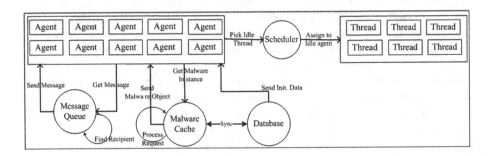

Fig. 1. MASFMMS architecture

Agent Pool consists of the agents that are waiting to be scheduled. These agents have been initialized as explained in Sect. 2.1.

Thread Pool consists of idle threads. These threads refer to machine threads that execute agent code. The threads on a machine being used for simulation are, typically, much less in number as compared to the agents.

Scheduler takes an agent from the agent pool and assigns it to a thread. The scheduler takes care that no agent is starved or scheduled more than its fair chance.

Message Queue receives messages from the agents and forwards them to their respective recipients. All agents register themselves with the message queue at the beginning of the simulation. Message Queue also allows broadcast messages. Hence the message queue serves as the network (See Sect. 2.2).

Malware Cache is a caching service provided by MASFMMS. This service accelerates the simulation by providing quick instances of malware objects, thus avoiding unnecessary database access to fetch malware descriptions for initialization. This is especially useful in the middle stages of proliferation when the infection rates are high.

3.2   Parameters of Simulation

Simulation wide parameters for MASFMMS are explained in Table 1. These parameters define the macro aspects of the simulation, such as its duration, properties of the network, the malware under investigation, the scanning limits of a computer. Basically these are the parameters that apply across the simulation. An important aspect to be noted here is *"Cycles"*. Cycle *"k"* is said to be complete when all agents have been scheduled k times and have finished execution. This definition is needed to provide balance to the simulation. In the beginning, the number of infected agents is less, and the simulation proceeds quickly as less processing is required. However, as the simulation continues and more and more agents become infected, each cycle takes more time. Hence *1 second* has no meaning in the simulation environment. In the real world, many instructions might get executed in *1 second*; in the simulation, a similarly large number of instructions will get executed in *1 cycle*. Hence the unit of time in MASFMMS is *Cycles*.

3.3   Agent State Diagram

Once the agents have been initialized, MASFMMS starts the simulation. The state transition diagram for the agents is shown in Fig. 2. Note that the two *"Processing Messages"* states are the same and are duplicated for clarity and ease of understanding. Along with the malware proliferation, the agent continues to respond to messages. Hence infection of an agent by multiple malware is possible. Also, the *"death"* of an agent means that it is removed from the agent pool and is no longer scheduled.

Table 1. Parameters of Simulation

Parameter	Symbol	Defination
Total Cycles	c	"c" is the maximum number of cycles
No. of Computers	N	The set of computers to be simulated by MASFMMS
Size of Hit List	H	The number of computers "seeded" with malware
Scanning Rate	S	The maximum requests that an agent can make in one cycle
Malware	V	The malware to be used in the simulation
Scanning Type	Q	The Function used by the malware for target selection
Category	T	
Infection Logic	-	Explained in Sect. 2.3
Post Infection Logic	-	
Patch	P_v	Patch "P_v" removes the vulnerability due to which malware V is able to infect
Patching Rate	p	The rate at which the vulnerability and malware are removed
Death Rate	d	The Rate at which the malware is removed without patching, leaving the computer vulnerable to re-infection

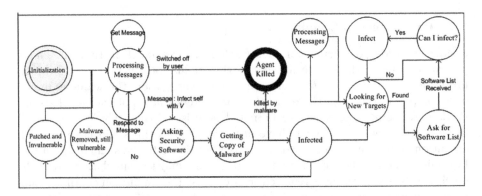

Fig. 2. State Transisition Diagram for Agents in MASFMMS

4   Experimental Results

4.1   Setup

In these experiments, we show how MASFMMS is used to simulate different kinds of security weaknesses and their exploits. We try out various different models and see how the results produced by the framework match their predictions.

The default senario used for simulation had about 100000 computers with 100 of them seeded with the malware. The malware, "Experimental Malware" had a threat level of "Category II" and used a "Randomized" scanning algorithm. It has a patch P_v with id "10". The network imposed a restriction of 100 scans/cycle and the death rate for this malware was 0.001, which is small and an acceptable value[9]. We did not include patching by default and ran the simulation for 50 cycles. This translates into the following values for the parameters (1)

c=50,(2) N=100000, (3) H=100, (4) S=100, (5) Q="Randomized",(6) d=0.001, (7) V="Experimental Malware", (8) T="Category II", (9) p=0, (10) P_v=10 (See Table 1). Any changes to these default values are specified during the experiment. In the experimental setup, all the agents have one program vulnerable to V. Some of the agents have security measures. The infection and post-infection logic vary with malware and are specified along with the experiment. We conducted experiments on a machine with four 3 GHz processors and 2GB of main memory using MySQL version 5.0.18 and Python version 2.3.

4.2   Worms

A computer worm is a self replicating computer program. It tries to deplete the resources of a computer, such as CPU, memory or network bandwidth. Many simulation models have been proposed for worms. For example Analytical Active Worm Propagation (AAWP), the classical epidemiological model, the two factor worm propagation model [9][3][5]. Due to the lack of space and to cover most features of the framework, we will see how MASFMMS can be used to simulate the Classical Epidemiological Model for worm propagation. This model assumes that computer worms proliferate in the same manner as biological viruses proliferate in the animal population [3] and has been used by [9][5] and others as a basis for comparison. The classical epidemiological model for finite population is described by $\frac{dn}{dx} = \beta n(1 - n) - dn$. The solution to this equation is $n(t) = \frac{n_0(1-\rho)}{n_0+(1-\rho-n_0)e^{-(\beta-d)t}}$, where $n(t)$ is the number of infected hosts at time t in a population of size N. The infection rate is β and the death rate is d. Also, $\rho = \frac{d}{\beta}$ and $n_0 \equiv n(t = 0) = \frac{sizeofhitlist}{N} = \frac{h}{N}$. We expected a sharp exponential rise in the infection due to the nature of the worm. Fig. 3 shows the results that were obtained, we see that the MASFMMS simulated worm is rising slower than predicted by the equation. This is due to the fact that differential equations are continuous whereas the simulator is discrete. [9] describes this variation in

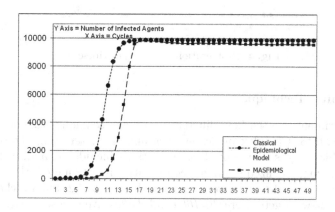

Fig. 3. Classical Epidemiological Model

detail. The graphs match the experimental results obtained by [10] using an email worm.

4.3   Trojan Horses

A Trojan horse is a malicious program that masquerades as a legitimate application or file. Trojan infections normally open a Back-door and/or "drop" a payload. This "dropped" payload is often a virus that infects the computer. For this experiment, we assume that both the Trojan and the payload of the malware follow a "worm" like infection pattern. This experiment shows how MASFMMS can be used to simulate the effect of multiple malware together. The infection logic for the payload will ask the agent to look for back-doors installed by the Trojan part of the malware. The infection logic for the Trojan would just look for a vulnerability, exploit it and install a back-door.

In this experiment, we took N=80000. We expected an initial rise of infection due to the Trojan horse (See Fig. 4). This infection leaves many back-doors. The payload worm then starts using these back-doors and starts infecting the agents. We could not find any experimental results for a similar experiment. However, the experiment described in Sect. 4.2 is similar to this one in its nature of infection rate, and gives similar infection graphs.

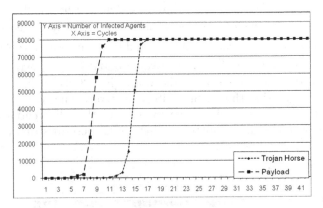

Fig. 4. Proliferation of a Trojan Horse

4.4   Scanning Techniques

In this experiment, we see how different scanning techniques can be modeled using MASFMMS. Complex malicious programs use interesting algorithms to locate possible targets. Some of these algorithms include techniques like randomized scanning, localized scanning, co-operative infection etc.[11]. In this experiment, we choose the same worm described in Sect. 4.2, but we change the way it selects new targets. Here, $h=1000$ and $d=0.0$. We choose linear, randomized and localized scanning because these are the most common types used by

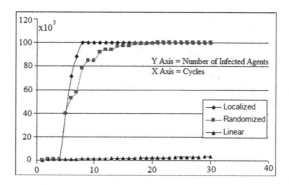

Fig. 5. Comparison of different scanning techniques used by the worms

malware. First, we try linear scanning. The worm starts generating ids of targets linearly from the beginning of the address space. [11] describes why this kind of technique is not used in practice. In the second run, we choose randomized scanning. The malware infection logic carries a random number generator that generates new targets. The advantage of this approach is that the malware is quickly scattered through the network, hence the scans themselves appear to come from everywhere. However, as the number of infections increases and the network starts to saturate, fewer probes reveal potential targets. In the third run, we use localized scanning. This algorithm uses the fact that vulnerable machines are often clustered together. The worm starts from the neighbors of the host and moves to agents which are further away. Linear scanning being a naive technique gave poor results. (See Fig. 5) However, randomized and localized scanning gave results which are similar to experimental results obtained by [11].

4.5 Effect of Procrastinating Patching

Patching a computer is defined as a process of removing a malicious program along with the security weakness described by a vulnerability from the computer. After patching, the computer becomes invulnerable to infection due to the exploitation of the said vulnerability.

Many users procrastinate the installation of patches on their computers once a patch is released. To simulate this scenario the procrastination equation $U = \frac{(E*V)}{(I*D)}$ [8] is used; U is the desire to complete the task, E is the expectation of success, V is the value of completion, I is the immediacy of the task and D is the personal sensitivity to delay. Based on this equation we define ten levels of procrastination(L). $L=10$ is for the most lazy users who procrastinate forever and $L=1$ is for the least lazy who patch the computer instantly. The malware used for this simulation is the worm described in Sect. 4.2 and the patch is released at $c=4$. As we can see in Fig. 6, the infection dies out very quickly for people with low L. The infection rises quickly and continues unabated for people with high L. We could not find supporting data for this experiment, but it shows

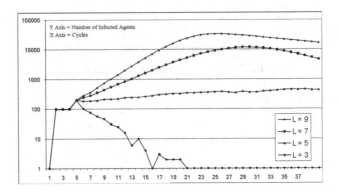

Fig. 6. Effect of Procrastination of patching on proliferation

the effectiveness of MASFMMS. We were able to introduce a new parameter L into the experiment and results for $L=9$ are very close to the results for the worm in Sect. 4.2.

5   Summary

In this paper, we developed a framework based on multi agent systems to simulate the effect and proliferation of malware across a computer network. We have tried to incorporate many parameters from taxonomy given by Weaver et al.[13]. Our framework has been implemented and the results show the viability of the approach. The advantages of our approach are seen in scalability in simulating a large number of computers, incorporating intangible parameters such as the behavior of humans (as agents) on computers, and the ability to incorporate a variety of malware parameters. The experiments that we have presented here show the diversity with which MASFMMS can be used, and now one can introduce new parameters in the simulation and dynamically control the agent and environment. The current development efforts are aimed at extending this framework to forecast the proliferation of malware across the network, and the impact of corrective actions taken to contain it.

References

1. CERT Coordination Center, http://www.cert.org/
2. CAIDA, http://www.caida.org/research/security/code-red/
3. Staniford, S., Paxson, V., Weaver, N.: How to Own the Internet in Your Spare Time. In: 11th Usenix Security Symposium, San Francisco (August 2002)
4. Frauenthal, J.C.: Mathematical Modeling in Epidemiology. Springer, New York (1980)
5. Zou, C.C., Gao, L., Gong, W.: Monitoring and early warning for internet worms
6. "NMAP", http://www.insecure.org/

7. Skaggs, B., Blackburn, B., Manes, G., et al.: Network vulnerability analysis. In: Proceedings of Wales E., Vulnerability assessment tools. Network Security, vol. 2003(7), pp. 15–17 (July 2003)
8. Steel, P.: The Nature of Procrastination: A Meta-Analytic and Theoretical Review of Quintessential Self-Regulatory Failure. Psychological Bulletin (2007)
9. Chen, Z., Gao, L., Kwiat, K.: Modeling the Spread of Active Worms. In: IEEE INFOCOM (2003)
10. Zou, C.C., Towsley, D., Gong, W.: Email Worm Modeling and Defense. In: 13th International Conference on Computer Communications and Networks (ICCCN 2004), Chicago, October 11-13 (2004)
11. Weaver, N.C.: Warhol worms: the potential for very fast Internet plagues (August 2001)
12. Nachenberg, C.: The Evolving Virus Threat. In: 23rd NISSC Proceedings, Baltimore, Maryland (2000)
13. Weaver, N., Paxson, V., Staniford, S., Cunningham, R.: A taxonomy of computer worms. In: Proceedings of the 2003 ACM workshop on Rapid malcode, Washington, DC, USA, October 27 (2003)

Towards a Formal Semantics of Event-Based Multi-agent Simulations

Jean-Pierre Müller

CIRAD, UPR GREEN, Montpellier, F-34398 France
Associate researcher to LIRMM, Montpellier, France

Abstract. The aim of this paper is to define a non-ambiguous operational semantics for event-based multi-agent modeling and simulation, applied to complex systems. A number of features common to most multi-agent systems have been retained: 1) agent proactive as well as reactive behavior, 2) *concurrency*: events can arrive simultaneously to an agent, an environment or any simulated entity and the actual change only depends on the target according to the influence/reaction paradigm [1], 3) *instantaneity*: if reaction takes time, perception as well as information diffusion is instantaneous and should be processed separately, 4) *structure dynamics*: the interaction structure (who is talking to whom) changes over time, and the agents as well as any simulated entity may be created or destroyed in the course of the simulation.

For each of these features, a solution inspired by the work on *DEVS* (Discrete EVent Systems, [2]) is proposed. *Proactive/reactive behavior* is naturally taken into account by *DEVS*. *Concurrency* is dealt with using *//–DEVS* (in [2]), a variant of the pure *DEVS*. *Instantaneity* is managed by distinguishing the physical events producing state transitions and the logical events realizing only perception and information diffusion. The *structure dynamics* is achieved by using a variant of ρ-*DEVS* (cf. [3]) where the expressiveness allows to manage hierarchical structures. The operational semantics is given as abstract algorithms and the expressive power of this formalism is illustrated on a simple example.

1 Introduction

The complex systems are characterized by a set of local components or entities in non-linear interaction whose global behavior is not reducible to any composition of the individual behaviors [4]. The question at the origin of the modeling process largely defines both the system and the entities to consider and therefore the point of view. The problem is even more complicated when it is necessary to articulate a set of these disciplinary points of view, possibly at various levels of organization [5,6], as it is often the case for eco-sociosystems. These disciplinary points of view produce a number of *thematic models* to be articulated. Many *formalisms* have been proposed to model the thematic models either at the aggregated level with differential equations, possibly partial for taking into account the spatiality of the phenomena, the cellular automata or, at the local

N. David and J.S. Sichmann (Eds.): MABS 2008, LNAI 5269, pp. 110–126, 2009.

(non-aggregated) level, individual or multi-agent based systems. An overview of the existing formalisms is given in [6]. Finally the models in these dedicated formalisms are implemented using raw programming languages up to specific simulation *platforms* like MatLab and Stella for dynamical systems, or Cormas [7] and Repast [8] for multi-agent systems, to cite a few. The modeling process going from the thematic model to its expression within a formalism down to the implementation platform needs to be made easier as more and more complex systems have to be modeled.

Our aim is to define an implementation platform such that any formalism can be mapped in a systematic (and then automatizable) way onto it. Among these formalisms, we shall concentrate on multi-agent systems. Theoretically, multi-agent systems are mainly characterized by the openness (entities coming in and out) and structure dynamics (the topology of interaction is dynamical). More than the definition of agents as having purposeful behavior (cf. [9]), we shall retain more generally the following features of the multi-agent based formalisms:

1. *reactive and proactive behavior*: an agents reacts to incoming events (if it wants to), and behaves proactively, regardless of the complexity of its internal architecture (from simple rules up to sophisticated reasoning),
2. *concurrency*: events can arrive simultaneously to an agent, an environment or any simulated entity. We consider that the actual change only depends on the target according to the influence/reaction paradigm [1], and that the result must not depend on the order of arrival nor on the order in which the entities are run as in many existing platforms,
3. *instantaneity*: if reaction takes (simulated) time, perception as well as information diffusion is instantaneous and should be processed separately,
4. *structure dynamics*: the interaction structure (who is talking to whom) changes over time and the agents as well as any simulated entity may be created or destroyed in the course of the simulation.

Moreover, we shall consider discrete event simulation in contrast with most existing multi-agent simulation (MAS) platform as Cormas [7], Repast [8] and many others which only consider fixed step simulations. The problem we want to tackle in this paper is to provide an implementation platform with a clear operational semantics, providing the above mentioned features, in which to map multi-agent systems, possibly combined with any other formalism.

Most, if not all, existing MAS platforms produce simulation results which do not depend only on the model but on the way the model is implemented and the scheduling ordered. This ordering is at worst arbitrary and at best randomized. We argue that it is an undesirable state of affairs and we propose to use a *DEVS*-inspired formalism to tackle this issue. *DEVS* was proposed by Zeigler [2] as a formalism to account for any kind of discrete event system. Since then a lot of *DEVS* extensions have already been proposed to handle concurrency [2] as well as structure dynamics. Among the later extensions, we can cite DS-*DEVS*[10], Dyn*DEVS*[11] and, more recently ρ-*DEVS*[3]. Some of these extensions have been proposed as tools for formalizing multi-agent systems [12,13]. However, as far as we know, no systematic account still exists. In this paper, we propose to address each of the desired features by proposing either an existing or a new extension of the *DEVS* formalism, showing how it can be used to take the feature into account. One of our main contribution to *DEVS* is a clear distinction between the physical transitions simulating time-dependent physical processes and the computation of the consequences which can be both structural and informational.

Each of the following sections shall present one of the retained features, using a simple example to illustrate the use of our proposed extensions. Finally, we shall present an abstract algorithm for all the proposed extensions before concluding.

2   The Example

To illustrate our proposal, we shall take the example of the firemen fighting against fire spreading in a landscape (inspired from [14]). We have to represent the landscape made of empty spaces and forests with its dynamics of fire occurrence and spreading. We assume that the fire occurs spontaneously with a given (low) probability. The firemen are wandering around in the landscape. If a fire occurs, they are informed of the direction relative to their position and they move towards the fire. When on a burning place, they water the surface.

The model is composed of three parts:

- S a space composed of cells C with either a Von Neuman (i.e. 4 neighbors) or a Moore (i.e. 8 neighbors) topology. Each cell can be either empty, with trees, watered, burning or burned. A cell with trees has a given probability to spontaneously burn;
- T a team of firemen F positioned on the cells and whose objective is to stop the fire;
- P be the position relation linking the cells of S to the firemen of T and reciprocally.

More than one team could be defined by extending P for handling several teams. Together, P, T and P define the global structure of the system we want to stimulate.

3   Reactive-Proactive Behavior and *DEVS*

DEVS [2] is a formalism based on discrete event simulation able to model:

- *atomic entities* with a set of events incoming and outgoing through input and output ports and a set S of internal states transiting in response to incoming events (reactive behavior) or spontaneously, after having sent outgoing events (proactive behavior) (see figure 1(a)), A *DEVS entity* is a tuple $< X, Y, S, \delta_{ext}, \delta_{int}, \lambda_{ext}, \tau >$ where:
  - X is a set of input events;
  - Y is a set of output events;
  - S is a set of states;

- $\delta_{ext} : Q \times X \to S$ is the *external transition function* implementing the reactive behavior, Q is S^+ composed of $s \in S$ and the duration since the last transition;
- $\delta_{int} : S \to S$ is the *internal transition function* implementing the proactive behavior;
- $\lambda_{ext} : S \to Y$ is the *external output function* only called before an internal transition;
- $\tau : S \to \Re^+$ is the *time advance function* giving the duration until the next output and then internal transition occurrence.

Notice that the output depends on the state before the internal transition and therefore is considered as a (delayed) answer to the previous state transition, mimicking delays in physical systems.

- *composed entities* recursively composed of entities coupled through their ports and with the composed entity input and output ports (see figure 1(b)). This compositionality is defined as a *closure under coupling* property of the *DEVS* formalism.

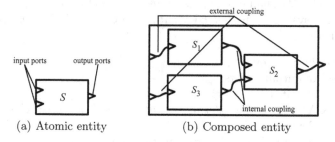

(a) Atomic entity   (b) Composed entity

Fig. 1. Atomic and composed entities

The operational semantics of an atomic entity is defined by an abstract algorithm embedded in a *simulator*. The operational semantics of a composed entity is defined by an abstract algorithm embedded in a *coordinator* which appears as a *DEVS* atomic entity for a recursively embedding composed entity. *DEVS* and any of its extensions must define the atomic entity structure, the composed entity structure and the related simulator and coordinator, obeying to the closure under coupling property. Although one can think of the simulation system as a tree of coordinators with simulators as leaves, the actual implementation often flattens the tree with a single coordinator in charge of a set of simulators. This single coordinator is then called a *DEVS-bus*.

A *DEVS*-bus computes the nearest date when an output and internal transition occurs using τ of each composing entity and then executes all the outputs, the internal transitions planned at that date and the consequent external transitions. The immediate limitation of the so-called pure *DEVS* is its arbitrary way to handle simultaneous events. When events are arriving simultaneously to the input ports of an entity, possibly simultaneous to an internal transition, the order of execution is controlled by a tie-breaking function defined in the coordinator.

3.1   Application to Multi-agent Systems

From an agent perspective, δ_{int} accounts for the proactive behavior whose occurrence is autonomously decided by the agent (τ), δ_{ext} accounts for its reactive behavior. In all cases, it is up to the agent to decide how to manage a potential conflict among the incoming events. No hypothesis is made on how these functions are computed. In particular, they could be simple condition/action rules up to sophisticated BDI reasoning, including probabilistic algorithms or not. Moreover, an agent is not necessarily a single *DEVS* entity but could be a composition of them where some are devoted to perception, others to action, etc.

4   Concurrency and //–DEVS

4.1   Introducing //–DEVS

//–DEVS [2] is an extension of *DEVS* where simultaneous events are transmitted together to the entity and the co-occurrence with an internal transition is signaled specifically. We shall briefly present *//–DEVS* before arguing its usefulness for multi-agent systems. An *//–DEVS entity* is a tuple $< X, Y, S, \delta_{ext}, \delta_{int}, \delta_{con}, \lambda_{ext}, \tau >$ where:

- X is a set of input events;
- Y is a set of output events;
- S is a set of states;
- $\delta_{ext} : Q \times X^b \to S$ is the *external transition function* where X^b is the set of bags over elements in X;
- $\delta_{int} : S \to S$ is the *internal transition function*;
- $\delta_{con} : S \times X^b \to S$ is the *confluent transition function*, subject to $\delta_{con}(s, \phi) = \delta_{int}(s)$;
- $\lambda_{ext} : S \to Y^b$ is the *external output function*;
- $\tau : S \to \Re^+$ is the *time advance function*.

As in *DEVS*, τ tells the simulator and thus the coordinator the duration before an internal (proactive) transition shall occur. When the duration elapses, a bag of output events (Y^b) – throughout the paper, the upper b stands for a bag, i.e. a set of elements with duplicates – is produced by λ_{ext} – called simply λ by the *DEVS* literates – and then the internal transition occurs (δ_{int}). A bag of external events (X^b) can occur anytime before $\tau(s)$ elapsed, producing an external transition (δ_{ext}) which depends on the dynamical state in Q. Finally, if the arrival of a bag of input events occurs exactly when $\tau(s)$ elapsed, the confluent transition function(δ_{con}) is called.

The coordinator is in charge to iteratively:

1. ask to each simulator the duration τ until the next internal transition,
2. let *IMM* be the set of simulators with the same minimum duration,
3. advance the global time by that duration,
4. call λ_{ext} for each simulator in *IMM*,

5. decide for each simulator whether it should do an internal (in *IMM* with no incoming events), an external (only incoming events) or a confluent transition (both in *IMM* and with incoming events).

This algorithm guarantees that all the possible simultaneously incoming events when a simulator is in a given *state* shall be known and transmitted as a bag. However, if the duration to the next internal transition is 0 (i.e. $\tau(s)$ can be 0), the simulators can make several state transitions at the same *date*. In the next section we propose to manage when $\tau(s) = 0$ and when $\tau(s) \neq 0$ separately to distinguish between the simulation of physical transitions (which take time) and what is mere information propagation.

4.2 Application to Multi-agent Systems

In [1], we argue that action in multi-agent systems requires a special attention. The view of action as a state transition does not resist to the fact that the environment or the other agents could actually not perform the expected state transition, nor does it resist to concurrency, i.e. to the fact that the actual environment state change results from a combination of the simultaneous actions by the agents. In [1], we propose to consider agent actions as influences of which effect depends on the receiver (the environment or another agent), in the so-called influence/reaction paradigm. The mapping of this idea into *//–DEVS* is immediate. The entities do not change directly the other entities states (would they represent an environment or another agent) but send events to them. From now on, we shall call these events, *influences* to stress this semantics. In addition, all the simultaneous influences are given at once to the receiving entity, delegating the actual state change and possible conflict resolution to the target of these influences. This property entirely complies with the influence/reaction philosophy. Moreover, no arbitrary choice is made by the coordinator, leaving to the modeler the entire responsibility of the model behavior. It contrasts with most existing MAS platforms where the choice is made arbitrarily by the scheduler, at best by randomizing the order in which the agents are run (e.g. the possibility to randomize is part of the comparison among MAS platforms in [15]).

In our example, we shall consider the case of the cellular automaton simulating fire spreading. At any state transition which are triggered at fixed time step, either by having $\forall s, \tau(s) = cst$ or by an external clock sending ticks to the cells, we chose each cell C to communicate its state to its neighbors. It means that any cell shall either perform an internal transition simultaneously to receiving up to four influences from its neighbors (δ_{con}) or perform an external transition with a tick and the same up to four influences (δ_{ext}). However if only an external transition is triggered, then no influences are output because the influences output only occur before an internal transition. Therefore only the solution where the state transition is internally triggered is possible, the influences from the firemen and the neighbors cells producing confluent transitions. According to the influence/reaction paradigm, the next state depends on all these influences at once: the advance of time for the probability of fire and the neighbors state and firemen actions for fire diffusion. Therefore, we argue that $//-DEVS$ naturally solves the influence/reaction problem.

5   Instantaneity and the Logical Influences

5.1   Introducing the Logical Influences

In the previous section, we stressed that all the simultaneous incoming influences shall be provided simultaneously to the target entity for any given state, but that several state transitions can occur at the same date if $\tau(s)$ can be 0. Clearly, a physical system never does repeated instantaneous transitions, so the question rather is the meaning of these instantaneous transitions in the DEVS implementation of the physical system simulation. We propose to illustrations of this point, one in the context of formal integration and another in the context of multi-agent systems.

In the context of formal integration, one could see a composed entity as a set of coupled differential equations, on per entity. However the value of the variable of each equation should be instantaneously known by all the differential equations it is coupled to. There are several solutions to this problem: 1) the equation solving process is driven by the transmission of the variable values; 2) the transmission of the variable value is intertwined with the step-wise solving process, i.e. one step for solving and one step for value transmission; 3) the use of $\tau(s) = 0$ for value transmission and $\tau(s) \neq 0$ for step-wise equation solving.

The same is true for observation in general and perception in particular. An agent must perceive its environment. In the $DEVS$ context, either the environment send influences to the agent just after each state change, or the agents send an influence for requesting information from another agent or the environment and could get an answer to his request – In a strict asynchronous communication semantics, no answer is necessarily expected, keeping agent autonomy –. In both cases, it could be made by passing time as in the solution 1 or 2 of the differential equations case, or by making $\tau(s) = 0$ for information transmission. In the first case, we consider perception as a physical process which takes time as any real physical process (a measurement device never responds instantaneously). In the second case, information diffusion, perception, or observation are considered timeless.

Consequently, we argue that $\tau(s) = 0$ is an implementation artifact to manage information diffusion, observation and diffusion and we propose to separately handle that case. Technically, we propose: 1) to forbid $\tau(s) = 0$ for the simulation of the physical transitions; 2) to add another mechanism based on *logical influences* for information diffusion. By contrast the influences producing physical state transitions shall be called the *physical influences*.

We define a $M-DEVS^{V1}$ *entity* as a tuple

$$< X, Y, S, \delta_{ext}, \delta_{int}, \delta_{con}, \lambda_{ext}, \tau, \delta_{log}, \lambda_{log} >$$

with the same definitions as $//-DEVS$ plus:

- $\delta_{log} : Q \to S$ the *logical transition function*;
- $\lambda_{log} : Q \to Y^b$ the *logical output function*.

and τ defined on $\Re^+ - \{0\}$ instead of $\Re^+$.

The semantics is identical to the $//{-}DEVS$ entity semantics but for the two additional functions. λ_{log} is called after each transition (including δ_{log}) to propagate the information about the new state, while λ_{ext} is only called before an internal transition. δ_{log} is used to make computations in response to the propagated information, changing the part of S which is just a consequence of the actual physical transition and possibly producing further information diffusion. Both functions depends on Q and they do not change the duration since the last physical transition (of course). No hypothesis is made on the order in which information is propagated unlike the physical influences. However, we keep the property that all the logical influences which must arrive simultaneously when an entity is in a given state are actually transmitted simultaneously, preserving state consistency. With this new semantics, a timeless asynchronous computation process is added to the time-dependent simulation process.

When looking again at our cellular automaton example, there is clearly two different logics:

1. a physical logics based on the time steps and the influences from the firemen,
2. an informational logics of propagation of the consequences: perception of the cell state for the firemen (possibly after a request), information about the neighbor's states for the cells.

We propose to use τ, λ_{ext} and δ_{ext} for the first one, and λ_{log} and δ_{log} for the second. Although reducible to $//{-}DEVS$, $M{-}DEVS^{V1}$ introduces a clearer separation of concerns. It shall further show its expressive power when combined with structure dynamics.

5.2   Application to MAS Simulations

Many MAS platforms are written in an object-oriented language like Java[8] or Smalltalk[7] and the agent and environment behaviors are very often directly written in the corresponding language as, for example, in Cormas, Repast or Mason. Therefore, the method call mechanism is automatically provided for propagating the consequences with two main limitations. The first limitation is technical: the method call is synchronous. Although the caller could ignore the result, it closes the door to parallel implementations. The second limitation is semantical: these method calls are completely free, letting the programmer change the state of any entity in the simulated system, regardless of the time coherence. In contrast, our proposal provides a cleaner semantics preserving the system coherence with respect to time management.

6   Structure Dynamics

The most important features of multi-agent systems are:

1. the openness: the agents can appear and die dynamically during the simulation;
2. the topological dynamics: the neighborhood changes over time as a consequence of the mobility of the agents in their environment (being social or physical), hence the topology of interactions.

These features make the huge difference between multi-agent systems and a set of coupled differential equations, for example, where the coupling is fixed and the equations cannot appear and disappear. We shall collectively refer to these features as structure dynamics.

For dealing with these properties, a number of extensions to $DEVS$ have been proposed. Barros in [10,16] proposes to add to each coupled model, a specific atomic entity called the *executive model*, whose state defines the set of composing entities and their topology. Uhrmacher allows in [11] each entity to specify the topology of the network in a formalism called DynDEVS, a proposal we shall follow. Notice that at that stage the closure under coupling property is no longer fulfilled and therefore, the formalism is no longer a $DEVS$ extension but more a $DEVS$-inspired formalism.

In the following, we shall introduce our proposal to manage the structural changes. Thereafter, we shall introduce a last, purely cosmetic, definition before providing the algorithm.

6.1   The $M{-}DEVS$ Entity

We define a $M{-}DEVS$ *entity* as a tuple

$$< X, Y, S, \delta_{ext}, \delta_{int}, \delta_{con} \delta_{log}, \lambda_{ext}, \lambda_{int}, \lambda_{log}, \lambda_{str} >$$

with the same definitions as $M{-}DEVS^{V1}$ plus:

- $\lambda_{int} : S \to (\Re^{+} - \{0\}) \times I$ which combines τ with the specification of what to do as an *internal influence* ($\in I$);
- $\lambda_{str} : S \to Y^{b}$ the *structural output function* producing structural influences.

The transformation of τ into λ_{int} is only cosmetic. The rational is to have a function to express not only when to do something τ but also what to do (something which could be encoded into S). Consequently, the modified $\delta_{int} : S \times I \to S$ depends on both the state and the internal influence.

This extension is entirely reducible to the previously defined models but facilitates the expression of the algorithms as well as making the formal notations symmetrical regarding the influence kinds. In consequence:

- λ_{int} and δ_{int} are in charge of the internal transitions using internal influences and therefore of the proactive behavior;

- λ_{ext} and δ_{ext} are in charge of the external transitions using the physical influences and therefore of the reactive behavior;
- δ_{con} is managing the simultaneity of an internal and external transition;
- λ_{log} and δ_{log} are in charge of the logical transitions using the logical influences and therefore the information diffusion;
- λ_{str} is in charge of the structural changes performed externally by the coordinator itself and using the structural influences.

Once again, the operational semantics is similar to the $M-DEVS^{V1}$ entity but the addition of the λ_{str} function which is called after each transition in the same way as λ_{log}. In fact these two functions are closely linked to one another because the consequences of a physical transition can be both informational through the λ_{log} function and structural through the λ_{str} function. Conversely, the structural changes may require to propagate information, in particular to initialize the newly created structures.

There is no δ_{str} function because the structural influences do not change the internal state of the entities but only creates, destroys and changes the topology among the entities. Let $\Sigma = (M, Z)$ be the simulation structure where M is the set of entities and Z encodes the topology of M as a function: $< d, p_o > \rightarrow \{< r, p_i >\}$ where d, resp. r, is the sender, resp. receiver, entity, p_o, resp. p_i, is the output, resp. input, port. We define a function $change : X^b \times \Sigma \rightarrow \Sigma'$ which, given a bag of structural influences x^b and a simulation structure $\Sigma = (M, Z)$, computes a new simulation structure $\Sigma' = (M', Z')$. In order to be consistent with the influence/reaction paradigm, Σ' must be uniquely defined and independent of the order in which the structural influences are considered.

We distinguish four types of structural influences:

1. creation: the influence creates a new $M-DEVS$ entity and attaches it to a port;
2. linkage: the influence connects the port of an entity to the entities designated by the port of another entity;
3. removal: the influence disconnects the entities linked to a given port (without destroying them);
4. deletion: the influence destroys the entity (and all the related port connections).

Whether these operations are only applicable to the issuing entity or not is open. In our current implementation, the first and last are only allowed on the issuing entity but the linkage and removal can be made anywhere as long as there is a path of successive ports (called a *port reference*) from the issuing entity.

6.2 Application to MAS Simulation

In parallel with the comparison with other MAS platforms in section 5, most existing platforms provide the primitives for creating new agents or objects as well as changing the topology. However, most of the time, these primitives are executed immediately, letting the programmer caring about the consistency with ongoing event propagation and the order in which the agents are run. In our case, all the structural changes are deferred after the physical transitions have been carried out and these changes are carefully ordered in our implementation. It guarantees that the resulting simulation structure does not depend on the order in which the modifications are issued.

The cellular automaton S as well as the team T are examples of the use of the creation structural influences. The whole structure (i.e. a grid of cells and a population of firemen interconnected together) as illustrated in figure 2 could be created "by hand" but it would be fairly cumbersome, generated automatically from a MAS specification ("compile time" generation), or generated dynamically by having S and T creating the cells and firemen when starting the simulation ("execution time" generation). In the latter case, it is enough to have λ_{str} in S and T generating one creation structural influence for each cell and each fireman, attaching each cell and fireman to a port of S and T respectively. Of course any number of teams can be put on the cellular automata by adding further Ts using the same P to manage collisions (if two firemen cannot be at the same place). Furthermore, a team could be embedded in an arbitrary number of environments with different topologies and semantics by adding further Ss and Ps.

Let "fireman" be a port of each cell for communicating with the fireman situated on the the cell and conversely let "cel" be the port for communicating with the cell the fireman is on (as illustrated in the figure 2 for the fireman F_6), any movement should change the connections of these ports to reflect the new

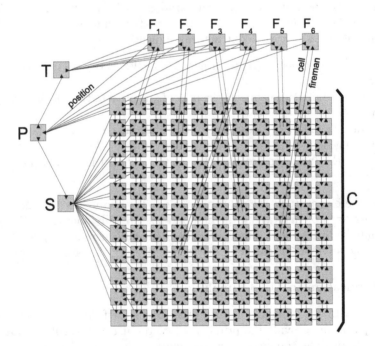

Fig. 2. The DEVS simulation structure of the example

situation. We propose to have a position P (see section 2) entity having ports to S and T respectively and performing the change using the structural influences each time a fireman moves on the grid. Accordingly, each fireman sends an influence to the position P for moving and P changes the connections among the fireman and its preceding and new cell.

Finally, when a fireman is surrounded by fire, it can also die by issuing a structural influence to delete it from the simulation. It can be done by issuing an influence to the position P to remove it from the current cell and another influence to T to remove it from the population.

Although, we provide a solution to this specific example, the mechanism is fully general and the structure could be automatically generated from higher level constructs.

7   The Abstract Algorithms

Given the various proposed extensions, it remains to formally specify the operational semantics of these constructs. As described in section 3, the operational semantics is defined by giving the abstract algorithm for the simulator (in charge of a single entity) and the coordinator (in charge of the whole simulation structure).

7.1   The M–$DEVS$ Simulator

The abstract algorithm of the simulator of such a model defines the answer to five messages (instead of three for $DEVS$, see [2]). For all the messages, the variable **parent** is assumed to designate the coordinator to which the simulator is attached:

- i-message(t) to initialize the model (algorithm 1): the results are yl^{b1} and ys^b the bag of logical and structural influences to possibly build the related structure and propagate information through it. These variables of the coordinator are set as a side effect of this algorithm.

Algorithm 1. initialization of an entity

$s = s_0$ {initial state}
$t_l = t$ {initial time}
$parent.yl^b = \lambda_{log}(s), parent.ys^b = \lambda_{str}(s)$

- *-message(t) to compute its output (algorithm 2): the result is the bag of the physical influences to send: ye^b;

Algorithm 2. the output of the model

Require: $t = t_n$
$parent.ye^b = \lambda_{ext}(s)$

---
[1] We are using upper b as the bag notation as in the section 4.

- x-message(x^b, t) to compute its state transition (algorithm 3): the results are the set of logical and structural influences yl^b and ys^b as a consequence of the state change;

Algorithm 3. computation of the physical state transition

Require: $t_l \leq t \leq t_n$
  if $t = t_n \wedge x^b = \emptyset$ **then**
    $s = \delta_{int}(s)$
  else if $t = t_n \wedge x^b \neq \emptyset$ **then**
    $s = \delta_{con}(s, x^b)$
  else if $t < t_n \wedge x^b \neq \emptyset$ **then**
    $s = \delta_{ext}((s, t - t_l), x^b)$
  end if
  $parent.yl^b = \lambda_{log}(s), parent.ys^b = \lambda_{str}(s)$

- l-message(x^b, t) to compute its logical transition (algorithm 4): the results are again further logical and structural influences yl^b and ys^b.

Algorithm 4. computation of the logical state transition

Require: $t_l \leq t \leq t_n$
  $s = \delta_{log}((s, t - t_l), x^b)$
  $parent.yl^b = \lambda_{log}(s), parent.ys^b = \lambda_{str}(s)$

- finally n-message(t) to set the current date t_l, to compute the next date t_n and internal influence yi (algorithm 5).

Algorithm 5. computation of the next internal transition

  $d, parent.yi = \lambda_{int}(s)$
  $t_l = t, t_n = t_l + d$

The output influences are assumed to be of the form: $< d, p_o, y >$ where d id the entity itself, p_o is the output port and y is the influence itself. For the structural influences only $< d, y >$ is necessary.

7.2 The M–$DEVS$ Coordinator

Normally in the $DEVS$ philosophy, the coordinator should manage a set of coupled M–$DEVS$ simulators and produce compositionally the same interface as a $DEVS$ entity. In this paper, we limit ourselves to a central coordinator (a $DEVS$-bus) taking in charge the complete structure $\Sigma = \{M, Z\}$ and shall describe this algorithm in several steps:

1. given M the set of simulators in Σ, the first step consists in getting the external influences of the entities about to simultaneously perform the next internal transitions (algorithm 6). Note that ye^b is set as a side effect of calling *-message(t).

Algorithm 6. Output before the internal transitions

$t = min_M(t_{n,d})$
$IMM = \{d | d \in M \wedge t_{n,d} = t\}, mail = \emptyset$
for $d \in IMM$ **do**
    send *-message(t) to M_d
    $mail = mail + ye^b$
end for

2. the second step consists in performing the actual physical transitions, collect-
   ing the logical and structural influences (algorithm 7). IMM shall contain
   both the candidates for internal transition and the receivers of external influences. Note that yl^b and ys^b are set as a side effect of calling x-message(t).

Algorithm 7. Execution of the physical transitions

$receivers = \{r | < d, p_o, y > \in mail \wedge < r, p_o > \in Z(d,p)\}$
$IMM = IMM + receivers, logical = \emptyset, structural = \emptyset$
for $r \in IMM$ **do**
    $x_r^b = \{x | < d, p_o, y > \in mail \wedge < r, p_o > \in Z(d,p)\}$
    send x-message(x_r^b, t) to r
    $logical = logical + yl^b$
    $structural = structural + ys^b$
end for

3. the third step consists in executing the structural changes, changing both M
   and Z (algorithm 8 in which *change* encodes the semantics of the structural
   influences as described in 6.1). The newly created simulators are initialized.
   In 8, yl_r^b and ys_r^b designate the bag of logical influences, respectively of
   structural influences, computed by the corresponding simulator r by calling
   their i-message . IMM shall contain the candidates for internal transition,
   the receivers of external influences and the newly created entities.

Algorithm 8. Structural changes

$M', Z' = change(structural, M, Z)$
$structural = \emptyset$
for $d \in M' - M$ **do**
    send i-message(t) to d
    $logical = logical + yl^b$
    $structural = structural + ys^b$
end for
$IMM = IMM \cup (M' - M), M = M', Z = Z'$

4. the logical influences are propagated (algorithm 9). yl^b and ys^b are defined
   as in 8 and computed when calling l-message. This step and the previous
   one are repeated iteratively until there is no structural or logical influences
   remaining.

Algorithm 9. Propagation of the logical influences

$list = logical, logical = \emptyset$
$receivers = \{r| < d, p_o, y >\in list \wedge < r, p_o, x >\in Z(d, p, y)\}$
for $r \in receivers$ **do**
    $x_r^b = \{x| < d, p_o, y >\in list \wedge < r, p_o, x >\in Z(d, p, y)\}$
    send **l-message**(x_r^b, t) to r
    $logical = logical + yl_r^b$
    $structural = structural + ys_r^b$
end for

Algorithm 10. Computes next internal transitions

---

for $d \in IMM$ **do**
    send **n-message**(t) to d
end for

---

5. IMM contains now all the simulators which incurred a physical transition or was newly created. These simulators are then asked to compute the date of their next internal transition for the next round (algorithm 10).

The use of a global structural *change* function hinders the possibility to distribute the model simulation. However, it is possible to overcome this limitation by composing recursively coupled *DEVS* entities into *DEVS* entities as illustrated in the figure 1(b). It is performed in three steps: 1) the introduction of the structure of a coupled *DEVS* entity as a set of *DEVS* entities and a number of input and output ports, 2) the extension of Z for coupling the input and output ports of the inner *DEVS* entities to the input and output ports of the coupled system, 3) the restriction of the structural influences in order to only modify the structure inside of the coupled system. One obtain a variant of *DEVS* in-between DS-DEVS[10] where a single *DEVS* entity is devoted to structural changes and ρ-*DEVS* [3] where all the entities can modify the structure including themselves.

8   Conclusion

This paper has presented $M-DEVS$ as a *DEVS*-inspired formalism for specifying the most important features of multi-agent systems, namely reactive and proactive behavior, concurrency, instantaneity and, most importantly, structure dynamics. A simple example has been used for illustrating the concepts. An operational semantics has been defined by providing an abstract algorithm to run models designed using $M-DEVS$. A clear distinction has been introduced between the physical transitions (using the internal and physical influences) and the computation of the consequences which can be both structural (using the structural influences) and informational (using the logical influences). These consequences are propagated until the overall structure stabilizes, before computing the date of the next physical transition.

Most, if not all, existing MAS platforms produce simulation results which do not depend only on the model but on the way the model is implemented and the scheduling ordered, this ordering being at worst arbitrary and at best randomized. We argue that it is an undesirable state of affairs and we propose a *DEVS*-inspired formalism with its algorithms which either delegates the potential ordering conflicts to the model, managing simultaneity, or orders the physical, logical and structural transitions such that the results do not depend on the issuing order, properly managing instantaneity and structure dynamics.

From the example, the resulting structure and dynamics reveal themselves as fairly complex and detailed. Actually, $M-DEVS$ should be thought as a kind of virtual machine specification for complex system simulation in which higher level specifications have to mapped. [2] has shown the possibility to map quantized dynamical systems into *DEVS*, others have mapped into *DEVS* or some of its extensions coupled differential equations and cellular automata [17]. We have illustrated the possibility to equally map multi-agent systems. In our case, an agent is mapped to a single *DEVS* entity. It could be further extended for complex agent architectures. For example, dealing with a variety of environments could require to separately specify various aspects of an agent to be coordinated, resulting in a composed architecture.

This proposal has been implemented and tested in the MIMOSA platform [18,19]. A number of applications are currently under development. From a theoretical point of view, a number of other extensions are under development like local time management, hierarchical coupling structures for holonic multi-agent systems and management of multiple points of view similar to the AGR approach [20]. The aim is to provide a formalism and its operational semantics for multi-level multi-agent systems.

I would like to thank Raphaël Duboz and my anonymous reviewers for helping me to clarify the issues, and to hopefully present it in a better way.

References

1. Ferber, J., Müller, J.-P.: Influences and reaction: a model of situated multiagent systems. In: Tokoro, M. (ed.) Proceedings of 2nd International Conference on Multi-Agent Systems, Kyoto, Japan, pp. 72–79. AAAI, Menlo Park (1996)
2. Zeigler, B.P., Kim, T.G., Praehofer, H.: Theory of Modeling and Simulation. Academic Press, London (2000)
3. Uhrmacher, A.M., Himmelspach, J., Röhl, M., Ewald, R.: Introducing variable ports and multi-couplings for cell biological modeling in devs. In: WSC 2006: Proceedings of the 37th conference on Winter simulation, Winter Simulation Conference, pp. 832–840 (2006)
4. Müller, J.-P.: Emergence of collective behaviour and problem solving. In: Omicini, A., Petta, P., Pitt, J. (eds.) ESAW 2003. LNCS, vol. 3071. Springer, Heidelberg (2004)
5. Müller, J.-P., Ratzé, C., Gillet, F., Stoffel, K.: Modeling and simulating hierarchies using an agent-based approach. In: Zerger, ndre., Argent, R.M. (eds.) MODSIM 2005 International Congress on Modelling and Simulation, Melbourne, Australia (December 2005)

6. Ratzé, C., Müller, J.-P., Gillet, F., Stoffel, K.: Simulation modelling ecological hierarchies in constructive dynamical systems. Ecological complexity 4, 13–25 (2007)
7. Le Page, C., Bousquet, F., Bakam, I., Baron, C.: Cormas: A multiagnet simulation toolkit to model natural and social dynamics at multiple scales. In: The ecoogy of scales, Wageningen, Netherlands (June 2000)
8. Tatara, E., North, M.J., Howe, T.R., Collier, N.T., Vos, J.R.: An introduction to repast modeling by using a simple predator-prey example. In: Proceedings of Agent 2006 Conference on Social Agents: Results and Prospects, Argonne, USA (2006)
9. Ferber, J.: Les systèmes multi-agents, vers une intelligence collective. InterEdition (1995)
10. Barros, F.J.: Modeling formalisms for dynamic structure systems. ACM Trans. Model. Comput. Simul. 7(4), 501–515 (1997)
11. Uhrmacher, A.M.: Dynamic structures in modeling and simulation: a reflective approach. ACM Trans. Model. Comput. Simul. 11(2), 206–232 (2001)
12. Uhrmacher, A.M.: Simulation for agent-oriented software engineering. In: Lunceford, W.H., Page, E. (eds.) First International Conference on Grand Challenges, San Diago, California (2003)
13. Duboz, R., Versmisse, D., Quesnel, G., Muzzy, A., Ramat, E.: Specification of dynamic structure discrete event multiagent systems. In: Proceedings of Agent Directed Simulation (Spring Simulation Multiconference), Hunstville, Alabama, USA (April 2006)
14. Hu, X., Muzy, A., Ntaimo, L.: A hybrid agent-cellular space modeling approach for fire spread and suppression simulation. In: WSC 2005: Proceedings of the 37th conference on Winter simulation, Winter Simulation Conference, pp. 248–255 (2005)
15. Railsback, S.F., Lytinen, S.L., Jackson, S.K.: Agent-based simulation platforms: Review and development recommendations. Simulation 82(9), 609–623 (2006)
16. Barros, F.J.: Abstract simulators for the dsde formalism. In: Medeiros, D.J., Watson, E.F., Carson, J.S., Manivanan, M.S. (eds.) WSC 1998: Proceedings of the 30th conference on Winter simulation, Los Alamitos, CA, USA, pp. 407–412. IEEE Computer Society Press, Los Alamitos (1998)
17. Wainer, G.A.: Modeling and simulation of complex systems with cell-devs. In: Proceedings of the 2004 Winter Simulation Conference, Washington DC, USA (December 2004)
18. Müller, J.P.: The mimosa generic modeling and simulation platform: the case of multi-agent systems. In: Coelho, H., Espinasse, B. (eds.) 5th Workshop on Agent-Based Simulation, Lisbon, Portugal, SCS, pp. 77–86 (May 2004)
19. http://sourceforge.net/projects/mimosa
20. Ferber, J., Gutknecht, O.: A meta-model for the analysis and design of organizations in multi-agent systems. In: Proceedings ICMAS 1998, Paris, France (1998)

A User Interface to Support Dialogue and Negotiation in Participatory Simulations

Eurico Vasconcelos[1], Jean-Pierre Briot[1,2], Marta Irving[3],
Simone Barbosa[1], and Vasco Furtado[4]

[1] PUC-Rio, Rua Marquês de São Vicente, 225
Gávea, Rio de Janeiro, RJ 22453-900, Brazil
jfilho@inf.puc-rio.br, jean-pierre.briot@lip6.fr, simone@inf.puc-rio.br
[2] LIP6, Université Paris 6 - CNRS
104 avenue du Président Kennedy, 75016 Paris, France
[3] UFRJ, Programa EICOS/IP, Av. Pasteur, Urca
Rio de Janeiro, 22290-240 RJ, Brazil
mirving@mandic.com.br
[4] UNIFOR, Av. Washington Soares, 1321
Edson Queiroz, Fortaleza, CE 60811-905, Brazil
vasco@unifor.br

Abstract. In this paper, we discuss the process of analysis and design of a user interface to support dialogue and negotiation between players of participatory simulations. The underlying context is an interdisciplinary project, named SimParc [8], about participatory management of protected areas for biodiversity conservation and social inclusion. This project is inspired by the ComMod MAS/RPG approach [6] and by recent proposals for software support for distributed role playing games, such as those by Guyot [14] and by Adamatti [1]. In this paper, we focus on the analysis and design of a user interface to ease and structure dialogue and negotiation between players, using a methodology based on semiotic engineering. Our main objective is to try to find a good balance between the necessary structuring and the spontaneity of dialog and negotiation.

1 Introduction

One of the principles of the Convention on Biological Diversity [15] refers to a participative process of social actors in the management of the biodiversity. Methodologies and computer-supported tools intending to facilitate this process have been addressed via bottom-up approaches that emphasize the role of local actors (stakeholders) and communities. Such bottom-up approaches echo the research conducted by members of the "ComMod" (for Companion Modeling) movement on participatory methods to support negotiation and decision-making for collective management of natural renewable resources. Their method, called MAS/RPG, consists in coupling multi-agent simulations (MAS) of the environment resources and role-playing games (RPG) by the stakeholders [6].

N. David and J.S. Sichmann (Eds.): MABS 2008, LNAI 5269, pp. 127–140, 2009.

Our project inherits from this tradition. It is named "SimParc" (which stands for "Simulation Participative de Parcs") and gathers French and Brazilian researchers in an inter-disciplinary approach. It constitutes an innovative and playful approach to explore and learn about negotiation procedures in national park management, based on the recognition of conflicts involving different interests, roles, and strategies. SimParc explores the use of advanced accompaniment methodologies based on MAS/RPG. More precisely, it follows recent proposals of integration of role playing into simulation, and of inserting artificial agents, as players [1] or as assistants [14]. In this paper, we focus on the user interface support for dialogue and negotiation between game players, and on the process of its analysis and design, by using a methodology for designing human computer interfaces based on a HCI theory named semiotic engineering.

2   The SimParc Project

2.1   Motivation

The SimParc project focuses on participatory management of parks and protected areas.[1] Our first concrete case study has been the urban National Park of Tijuca, in Rio de Janeiro, Brazil. It undergoes a real pressure, by urban growth and illegal occupation. This makes the question of the conflict resolution one of the key issues for the management of the park. Examples of inherent conflicts are: irregular occupation, inadequate tourist exploration, water pollution, degradation of the environment and illegal use of natural resources. Examples of social actors involved are: park managers, researchers, traditional or non traditional community representatives, tourist operators and agencies.

The design of our current role playing game has taken inspiration in real cases such as the National Park of Tijuca, in order to bring concrete elements to the game, which confers greater applicability to our proposal. However, we chose not to reproduce a real case but to simulate emblematic and illustrative real situations in national parks.

2.2   Objective

The SimParc game constitutes an innovative and playful approach to support negotiation procedures in national parks management. Current game has a pedagogical objective and is not (or at least not yet) aimed at decision support.[2] The targeted public includes managers of parks and protected areas, researchers, students, and all stakeholders and people willing to understand and explore the

---

[1] Parks are one among the different types of protected areas, as defined by Brazilian legislation. Other examples of types are, e.g., biosphere reserves or ecological stations [15].

[2] Current game is aimed at helping participants to discover and explore conflicts as well as negotiation strategies to address them. But we do not expect the resulting decisions to be directly applied to a specific park. This would require, e.g., a precise calibration and a predictive model for park viability. This is left for future work.

challenges, conflicts and process of negotiation for participative management of parks and protected areas.

The game is based on the process taking place within the council of the park. This council, consultative, includes representatives of various stakeholders (e.g., traditional community, tourism operator, environmentalist non governmental association, water public agency...). The actual game focuses on a discussion within the council about the demarcation ("zoning") of the park. More precisely, it is about the decision to associate a type of conservation (and therefore, use) to every sub-area[3] (named "landscape unit") of the park. We consider nine pre-defined potential types of conservation/use, from more restricted to more flexible, as defined by the law.

The game considers a certain number of players roles, each one representing a certain stakeholder. Depending on its profile and the presence of elements of concerns in some of the landscape units (e.g., tourism spot, people, endangered species...), each player will try to influence the decision about the type of conservation for each landscape unit. It is clear that conflicts of interest will quickly emerge, leading to various strategies of influence and negotiation (e.g., coalition formation, trading mutual support for respective objectives...).

The manager of the park observes the negotiation taking place and takes the final decision for types of conservation for each landscape unit, based on the legal framework, on the process of negotiation between players, and on his personal profile (e.g., more conservationist or more open to social concerns) [15]. He also may have to explain his decision, on demand from the players. The park manager may be played by a human, or by an artificial agent [9].

In summary, the objective of the project is thus to propose an epistemic process to help each participant discover and understand the various factors, conflicts, and the importance of dialogue for a good management of protected spaces.

2.3 Steps

The game is structured along six steps, as illustrated at Figure 1. At the beginning (step 1), each player is associated to a role. Then an initial scenario is presented to each player, including the setting of the landscape units, the possible types of use and the general objective associated to his role. Then (step 2), each player decides a first proposal of types of use for each landscape unit, based on his understanding of the objective of his role and on the initial setting. Once all done, proposals by players are made public to all. In step 3, players start interacting and negotiating about their proposals. This step is, for us, the most important, where players will collectively build their knowledge by means of argumentation process. In step 4, they revise and commit to their proposals. In step 5, the park

---

[3] We suppose that the process of identification of (or, decomposition into) the landscape units of the park has already taken place before. Actually, the two processes were considered simultaneously in an initial version of the game, but this proved too complex. Moreover, deciding the type of conservation alone is sufficiently effective to capture conflicts and negotiation between stakeholders.

manager makes the final decision, considering the process of negotiation, the final proposals and also his personal profile (e.g., more conservationist or more sensitive to social issues). Each player can then consult various indicators of his/her performance (e.g., closeness to his initial objective, degree of consensus, etc.). He can also ask for explanation of the park manager decision rationales. The last step (step 6), "closes" the cycle and provides a feedback on the decision, both by the players (indicating their level of acceptance of the decision) and some evaluation of the quality and of the decision through indicators (e.g., on the economical or social feasibility) or simulation.

A new cycle of negotiations may then possibly start (see Figure 1), thus creating a cycle similar to a learning cycle [17]. The main objective is indeed for participants to: understand the various factors and perspectives involved and how they are interrelated, negotiate to try to reach a group consensus, and understand cause-effect relations based on decisions.

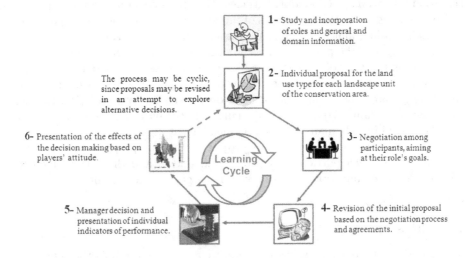

Fig. 1. The 6 steps of SimParc game

2.4   Game Computer Support

A computer support is proposed for the game, allowing distributed role playing, where each player acts and interacts via a computer interface, as has been pioneered by Simulación [14] and ViP-GMABS [1]. In SimParc, the role playing game is completely distributed and the master of the game will be automated, in part or completely. Because all interactions, decisions and actions are mediated by the computer, they can be formatted as objects, recorded and processed on-line or off-line to allow the management of the history of the negotiations (different ways of visualization of exchanged messages) and to evaluate and analyze the players and the negotiation process. In section 5, we will discuss the

Fig. 2. Test of the SimParc game version 1

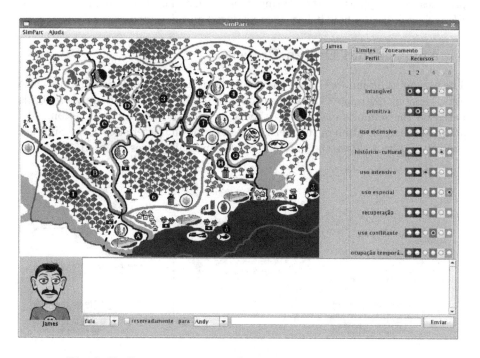

Fig. 3. SimParc game version 1 first computer-support prototype

interface support for the negotiation process. We are also working on introducing artificial agents in the game: (1) an automated park manager taking decision and explaining it, (2) artificial players [1], (3) assistants to the players [14], as discussed in [9] and in future publications.

2.5 Versions and Experiments

The initial design of the game (version 1) was conducted during year 2007. It was tested, without any computer support, through a game session conducted in September 2007 (see Figure 2). There were six roles in the scenario. Each role was played by a team of two players. Players were researchers and students of the APIS (Áreas Protegidas e Inclusão Social – Protected Spaces and Social Inclusion) research group, at UFRJ (Rio de Janeiro), led by Marta Irving, and specialized in biodiversity participatory management.

In parallel, a fist computer support prototype, based on the framework Simulación [14] was designed and built (see Figure 3).

Based on the evaluation of the first version of the game – notably via the analysis of the test of September 2007 – and on the evaluation of its computer-support prototype, we then designed a second version of the game, with a new computer support prototype under current construction. Among some specific features (artificial agents and automated evaluation of players performance), it provides some support for structuring interaction and negotiation among players. We will now discuss how it has been analyzed and designed.

3   Analysis and Design Process

The process of design was based on communication-centered design, and its more agile version, eXtreme Communication-Centered Design [3], design proposals based on the semiotic engineering theory of human-computer interaction.[4] We adapted the application of the methodology to the characteristics of the SimParc project. Figure 4 shows the different phases and sub-phases adopted.

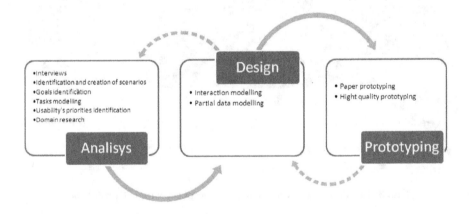

Fig. 4. The process of design

The output products of the analysis phase are the records from interviews with experts and users, scenarios (use cases), goals diagram and tasks model.

---

[4] According to it, both designers and users are interlocutors in an overall communication process that takes place through the interface of the system. Designers must tell users what they mean by the artifact they have created, and users must try to respond to what they are being told [22].

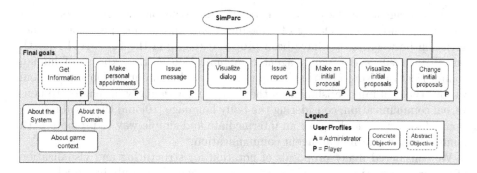

Fig. 5. The MoLIC diagram of final goals

The scenarios were constructed based on interviews, in a narrative form, to help at identify contextualized types of usages. The goals diagram (see Figure 5,[5] modeled in the MoLIC language (MoLIC stands for Interaction Modeling Language for the Conversation), was constructed from the scenarios and interviews, with the aim of representing the goals (identified a priori) of the users. We believe that the task model represents an intermediary step, easing a conceptual transition from the analysis phase (what, why and by whom) to the design (how). Note that task models are also widely used and accepted in human-computer interaction (HCI) [4]. Overall, the goal of the diagrammatic representation of task models is to provide an overview of the design process for each goal and how these goals are decomposed into tasks and sub-tasks. This diagram provides a new set of information about the process, presenting the hierarchy and flow of tasks, preparing designers and users to an outline of the interaction. We used an adaptation of the Hierarchical Task Analysis (HTA) [2] for modeling tasks identified from the goals diagram and the scenarios.

4 Design of an Interface Language for Negotiation Support

We consider negotiation as a particular form of communication process between two or more parties, focused on mutual agreement(s) on a given conflict of interest or opinions [18]. We further believe that the adoption of an interface language, based on argumentation models and linguistics theory, can offer different ways of support to a computer mediated negotiation process. The main objective for that interface language is to find the inflection point between the necessary "framing" and the maintenance of fluidity and naturalness of the dialogue.

The structure of the dialogue is an important factor, because it helps at a better management of the history of the negotiations facilitating the inclusion

---

[5] Because of the space limitation, we only illustrate some of the notations/diagrams. Figure 5 shows the final goals diagram. Note that there is also an instrumental goals diagram [5].

of artificial agents in the process, increasing the focus on the process, on issues negotiated and on the clarity of dialogue. Many interaction protocols for negotiation between agents have been proposed (e.g., via the FIPA-ACL effort), but they privilege the agent- agent communication at the expense of human communication. Note also that computer mediated communication suffers from various types of impoverishment of the dialogue, particularly in relation to nonverbal communication, considering the body language [10] and the vocal intonation. Thus, we are looking for an intermediate and simple way to promote both human-human and human-agent communication.

We considered many proposals of notation for structuring and visualization of the argumentation, as, e.g., in [16]. Among them: the Toulmin model, a reference for the majority of the posterior models; the Issue-Based Information System (IBIS), an informal model based on a grammar that defines the basic elements present in dialogues about decision-making; the "Questions, Options and Criteria" (QOC); the "Procedural Hierarchy of Issues" (PHI) and the "Decision Representation Language" (DRL) [16]. Based on this analysis, we believe that it is possible to offer a pre-structure, adding to the informal and interpretative characteristic of prose, while maintaining the fluidity of dialogue. Our main inspirations for rhetorical markers is IBIS [16], as well as theories of negotiation, e.g., [23] [19] and Speech Act Theory [21]. These markers are basically composed of rhetorical identifiers of intention (see Figure 6), the object focus of the intention and of a free speech (see Figures 8.3 and 8.4). These elements give the tone of the dialogue, making clear the illocution, and thus facilitating the expression of the desired perlocution [21].

We therefore provide the structure by threading from the dialogue, which minimizes risks of losing context, common in computer mediated communication (via chat) [12]. Figure 7 shows an example of threading based on the proposed structure.

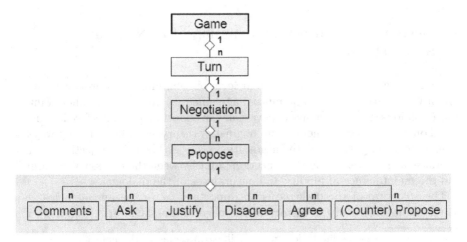

Fig. 6. Semi-structure for the text based on rhetorical markers

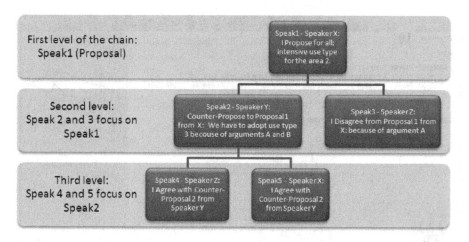

Fig. 7. Example of threading structured by the rhetorical markers

In complement to this semi-structure applied to the text, we propose to model each speak from players as an object. These objects have the following attributes: identifier, sender, receiver(s), marker, focus, and a free text (See Figure 7). This modeling eases at the management and indexing of dialogue by the system. For instance, filters may be applied to analyze the history of a dialogue, e.g., filtered along a given speaker, or a specific type of marker. But it also opens the way for its processing by software agents.

5 Prototype

The outputs of the design phase are: interaction diagrams, class diagrams, class and entity relationship model for the database. We then created a fast prototype in order to evaluate the appearance and usage. In the following we focus on the prototype interface corresponding to step 3 of the game, i.e., negotiation between players. It is indeed a central part of the game, when the shared knowledge is jointly negotiated and built. We would like to emphasize that we try to balance a support for some structure of the text of the dialogue and also sufficient fluidity.

The prototype user interface (see Figure 8) includes an area (Fig. 8.1) for the history of messages exchanged. The area (Fig. 8.2) for managing the history of messages offers different ways of selecting and ordering the information and includes a simple way to better identify speakers (discrimination by color). The area (Fig. 8.3) contains options for semi-structure of messages via rhetorical markers for intention (e.g., disagree). The area (Fig. 8.4) is for writing the actual contents (text) of the message. The area (Fig. 8.5) allows selecting the recipients (unique or multiple) of the message to be sent. The area (Fig. 8.6) provides the selection of iconic expressions to offer an alternative way for the user to express his emotional context during the negotiations, as an alternative way of minimizing the loss of communication modalities. The "facecons" were produced from

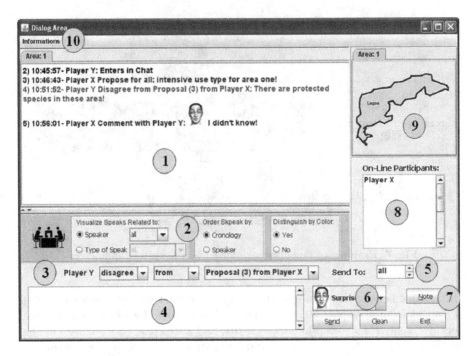

Fig. 8. Prototype interface for the negotiation step

the tool Artnatomia [11] which generates iconic facial expressions of emotional states from the virtual manipulation of the muscles of the face.

There is an area (Fig. 8.7) for personal annotations, allowing the user to make and record personal notes during the negotiation (this need was observed during test with players, see Figure 2). There is also an area (Fig. 8.8) with the list of participants and their roles that, in accordance with the C3 (Communication, Coordination and Cooperation) model [13], is a form of support for coordination. There is an area (Fig. 8.9) with the view of the object negotiated and its geo-processed representation. Last, menus (Fig. 8.10) are available with access to different types of information about the domain, the system and the context of the game, such as the legal types of land use, the roles, the game objective and phases, the system use and help.

6   Discussed and Open Issues

We now address some of the issues that were discussed during the workshop as well as issues that we encounter during our project.

6.1   Scope and Realism of the Simulation

Our objective is social simulation, based on a role playing game, including human actors playing some roles in a simulated situation of conflicts, negotiation and decision making. Our scenario for the role playing game has taken inspiration in real cases, such as the National Park of Tijuca, although it is not the reproduction of a real case. Real cases are important, because they bring concrete elements to the game, which allows our proposal to be evaluated in more realistic and illustrative settings. However, we chose not to reproduce exactly a real case, in order to leave the door open for broader game possibilities [9] [15].

The last step (step 6) of the game (see Figure 1) "closes" the cycle (and the epistemic loop) by considering the possible effects of the decision. In the current game, the players provide a simple feedback on the decision by indicating their level of acceptance of the decision. For a future version, we also plan to introduce some evaluation of the quality of the decision through computable indicators (e.g., on the economical or social feasibility). An alternative is a multi-agent simulation of the evolution of resources. We have not yet addressed this additional (and more traditional) type of simulation, because our current project main focus is on social simulation, negotiation and decision support. We are also aware that modeling the resources and their evolution (flora, fauna, water, animals, population and their activities) is always a challenge. It also triggers the issue of realism and predictiveness. In our project current stage, we are concerned with credibility and not yet with realism because our objective is epistemic and not about producing an (hypothetical) optimal decision.

6.2   Towards Hybrid Simulation

As already mentioned in Section 2.4, we are planning to introduce artificial players into the game [9]. The idea is to possibly replace some of the human players by artificial players (artificial agents). The social simulation will therefore become hybrid, with human and artificial agents in the simulation. A first motivation is to address the possible absence of sufficient number of human players for a game session [1]. But this will also allow more systematic experiments about specific configurations of players profiles, because of artificial players objective, deterministic and reproductible behaviors.

6.3   Hybrid Negotiations

An important and difficult issue for negotiation models is to reach some balance between human players needs and artificial players requirements. Negotiation languages for human players is usually richer and/but also more ambiguous. Negotiation languages for artificial players are usually more restricted in order to be unambiguous and interpretable by machine. In our project, we are currently exploring in parallel the two dual ways: (1) structure human negotiations though language and interface support and observe them (see Section 4); (2) design artificial players/negotiations, and insert/test them (see [1] [9]). Our mid-term goal is to gradually better understand human negotiation and see how to find a compromise with automated negotiation requirements.

6.4   From Modeling to Simulation and from Simulation to Modeling

One of the key issue of computational modeling and simulation of complex phenomena is about extracting knowledge about the phenomena, in our case social actors and social processes. This means the elicitation of models of representation, models of interaction and models of decision. Traditional approach used in social sciences and in computational modeling and simulation of social processes use observation and transcription of social actors behaviors in the real world, by using an ethnographic approach and also surveys based on interviews.

An alternative (participatory simulation, see e.g., [14]) is to directly involve human social actors as elements of the computer supported simulation of this social process. Computer supported role playing games create simulated situations in which social actors are immersed, can play their roles and expose their behaviors and strategies. Indeed, role-playing games are "social laboratories", because players can try many possibilities, without real consequences [6]. This leads to a more natural incremental modeling of the social process and of the behaviors of the social actors.

We then may gradually replace human actors by artificial agents (see Section 6.2), the human actors validating or amending the behaviors of artificial players. These artificial players may be programmed at hand (see, e.g., in [1]) or inferred by automated analysis of the human players. Indeed, the fact that the role playing game is distributed and that players interact through computers allows the systematic memorization of all interactions and decisions taking place between players. This opens the way for some automated or semi-automated analysis of traces of interactions [14] [8], in order to infer behavioral models. This means that elicitation (knowledge extraction) of human experts behavioral models (e.g., models of interaction, decision and negotiation) may be conducted via automatic monitoring of experts in (virtual/simulated) situation/action, as opposed to more traditional interview-based (off situation) elicitation. We believe that this represents of some kind of "virtuous circle", where modeling and simulation incrementally reinforce each-other.

7   Conclusion

In this paper, we have presented and discussed the process of analysis and design of a prototype user interface to support dialogue and negotiation for participatory simulations, for the domain of protected areas management. An important objective for the interface is to explore some balance between structuring dialogue and negotiation, e.g., through rhetorical markers, and keeping some fluidity. Our current prototype is under completion and we will soon start to test it by organizing game sessions with players expert in the domain of the game. We are planning to use epistemic tools proposed by semiotic engineering (which was used for the analysis and design phases) to test and evaluate the acceptance and usability by users during game sessions. We also plan to study the possible generality of some of the principles of our prototype interface for other types and domains of participatory simulations and serious games.

Acknowledgments. We would like to thank Gustavo Melo, Altair Sancho and Ivan Bursztyn for their main contribution to the design of the SimParc game, Vinícius Sebba Patto for his design of the first version of the prototype and for his design of assistant agents, Diana Adamatti for his design of artificial players, Alessandro Sordoni for his design of the artificial manager, Davis Sansolo for his design of the GIS system, and Paul Guyot for his past participation to the project. This research is funded by: the ARCUS Program (French Ministry of Foreign Affairs, Rgion Ile-de-France and Brazil) and the MCT/CNPq/CT-INFO Grandes Desafiós Program. Some additional individual support is provided by Alban (Europe), CAPES and CNPq (Brazil) fellowship programs.

References

1. Adamatti, D.F., Sichman, J.S., Coelho, H.: Virtual Players: From Manual to Semi-Autonomous RPG. In: Proceedings of the International Modeling and Simulation Multiconference (IMSM 2007), Buenos Aires, Argentina (February 2007)
2. Annett, J., Duncan, K.D.: Task Analysis and Training Design. Journal of Occupational Psychology 41, 211–221 (1967)
3. Aureliano, V.C.O., Silva, B.S., Barbosa, S.D.J.: Extreme Designing: Binding Sketching to an Interaction Model in a Streamlined HCI Design Approach. In: VII Simpósio Brasileiro sobre Fatores Humanos em Sistemas Computacionais (IHC 2006), Natal, RN, Brazil (November 2006)
4. Barbosa, S.D.J., de Souza, C.S., Paula, M.G., Silveira, M.S.: Modelo de Interação como Ponte entre o Modelo de Tarefas e a Especificação da Interface. In: Anais do V Simpósio sobre Fatores Humanos em Sistemas Computacionais (IHC 2002), Fortaleza, CE, Brazil, October 2002, pp. 27–39 (2002)
5. Barbosa, S.D.J., Paula, M.G., Lucena, C.J.P.: Adopting a Communication-Centered Design Approach to Support Interdisciplinary Design Teams. In: Proceedings of Bridging the Gaps II: Bridging the Gaps Between Software Engineering and Human-Computer Interaction, Workshop at the International Conference of Software Engineering (ICSE 2004), Scotland (May 2004)
6. Barreteau, O.: The Joint Use of Role-Playing Games and Models Regarding Negotiation Processes: Characterization of Associations. Journal of Artificial Societies and Social Simulation 6(2) (2003)
7. Bordini, R.H., Hubner, J., Wooldridge, M.: Programming Multi-Agent Systems in AgentSpeak using Jason. Series in Agen Technology. Wiley, Chichester (2007)
8. Briot, J.-P., Guyot, P., Irving, M.: Participatory Simulation for Collective Management of Protected Areas for Biodiversity Conservation and Social Inclusion. In: Proceedings of the International Modeling and Simulation Multiconference 2007 (IMSM 2007), Buenos Aires, Argentina, pp. 183–188 (February 2007)
9. Briot, J.-P., Vasconcelos, E., Adamatti, D., Sebba, V., Irving, M., Barbosa, S., Furtado, V., Lucena, C.: A Computer-based Support for Participatory Management of Protected Areas: The SimParc Project. In: Anais do XXVIII Congresso da Sociedade Brasileira de Computação (CSBC 2008) - Seminário Integrado de Software e Hardware "Grandes Desafios", Belem, PA, Brazil, pp. 1–15 (July 2008)
10. Ekman, P.: Emotions Revealed. Times Books (US), Weidenfeld & Nicolson Ekman, New York, London (2003)
11. Flores, V.C.: Artnatomia (2005), http://www.artnatomia.net

12. Fuks, H., Pimentel, M., Lucena, C.J.P.: R-U-Typing-2-Me? Evolving a Chat Tool to Increase Understanding in Learning Activities. International Journal of Computer-Supported Collaborative Learning (ijCSCL) (2006)
13. Gerosa, M.A., Pimentel, M., Fuks, H., de Lucena, C.J.P.: Development of groupware based on the 3C collaboration model and component technology. In: Dimitriadis, Y.A., Zigurs, I., Gómez-Sánchez, E. (eds.) CRIWG 2006. LNCS, vol. 4154, pp. 302–309. Springer, Heidelberg (2006)
14. Guyot, P., Honiden, S.: Agent-Based Participatory Simulations: Merging Multi-Agent Systems and Role-Playing Games. Journal of Artificial Societies and Social Simulation 9(4) (2006)
15. Irving, M.A. (org.): Áreas Protegidas e Inclusão Social: Construindo Novos Significados. Aquarius, Rio de Janeiro (2006)
16. Kirschner, P.A., Shum, J.B., Carr, S.C. (eds.): Visualizing Argumentation: Software Tools for Collaborative and Educational Sense-Making. Springer, Heidelberg (2003)
17. Kolb, D.A.: Experimental Learning: Experience as the Source of Learning and Development. Prentice-Hall, Englewood Cliffs (1984)
18. Putnam, L.L., Roloff, M.E. (eds.): Communication and Negotiation. Sage Annual Review Series, vol. 20. Sage, Newbury Park (1992)
19. Raiffa, H.: The Art & Science of Negotiation. Harvard University Press, Cambridge (1982)
20. Rao, A.S.: AgentSpeak(L): BDI Agents Speak out in a Logical Computable Language. In: Perram, J., Van de Velde, W. (eds.) MAAMAW 1996. LNCS, vol. 1038, pp. 42–55. Springer, Heidelberg (1996)
21. Searle, J.R.: Speech Acts: An Essay in the Philosophy of Language. Cambridge University Press, Cambridge (1969)
22. de Souza, C.S.: The Semiotic Engineering of Human-Computer Interaction. MIT Press, Cambridge (2005)
23. Wall, J.A., Blum, M.W.: Negotiations. Journal of Management 17(2), 273–303 (1991)

Towards Agents for Policy Making

Frank Dignum[1], Virginia Dignum[1], and Catholijn M. Jonker[2]

[1] Dept. Information and Computing Sciences, Utrecht University, The Netherlands
{dignum,virginia}@cs.uu.nl
[2] Technical University Delft, The Netherlands
C.M.Jonker@tudelft.nl

Abstract. The process of introducing new public policies is a complex one in the sense that the behavior of society at the macro-level depends directly on the individual behavior of the people in that society and ongoing dynamics of the environment. It is at the micro-level that change is initiated, that policies effectively change the behavior of individuals. Since macro-models do not suffice, science has turned to develop and study agent-based simulations, i.e., micro-level models. In correspondence with the good scientific practice of parsimony, current ABSS models are based on agents with simple cognitive capabilities. However, the societies being modeled in policy making relate to real people with real needs and personalities, often of a multi-cultural composition. Those circumstances require the agents to be diversified to accommodate these facts.

In this positioning paper, we propose an incrementally complex model for agent reasoning that can describe the influence of policies or comparable external influences on the behavior of agents. Starting from the BDI model for agent reasoning, we discuss the effect when personality and Maslow's hierarchy of needs are added to the model. Finally, we extend the model with a component that captures the cultural background and normative constitution of the agent.

In the paper we show how these extensions affect the filtering of the desires and intentions of the agent and the willingness of the agent to modify its behavior in face of a new policy. This way, simulations can be made that support the differentiation of behaviors in multi-cultural societies, and thus can be made to support policy makers in their decisions.

1 Introduction

Effective social simulation and effective support for policy makers depends on our ability to model the adaptive individual decision making process given subjective social norms, individual preferences, and policies. Where policy makers on different levels believe they only act successfully within bounds prescribed by social norms, they struggle to force big changes top-down if they dont seek social support. In fact, social norms have measurable consequences for the environment, e.g. energy intensive consumerism and lifestyles, have lead to the ecological near-crisis now at hand [26]. Moreover, Kable and Glimcher showed that social norms vary from culture to culture, by proving that the social discounting factor is not a unique scientific number, but a very subjective value differing from culture to culture [17]. Simultaneously, policies are based on aggregate top-down assumptions of economic behavior, whereas many changes occur bottom-up

N. David and J.S. Sichmann (Eds.): MABS 2008, LNAI 5269, pp. 141–153, 2009.

due to heterogeneity which is at the basis of cultural drift and adaptivity of a population. Due to heterogeneity, changes may occur that were not foreseen at the introduction of a policy, leading to a low efficacy of a policy, if anything is achieved at all. To support policy makers in their policy design, it is therefore desirable to evaluate proposed policies with models that are not based on economically desirable behavior (equilibrium theory), but on models that take realistic social interaction and cultural heterogeneity into account. Models to evaluate policies, should consequentially include complex agents that more realistically mimic human adaptive behavior. Also the environment in which the agents act needs to be represented dynamically, under the assumption that the environment influences the behavior of agents as well. Because of the magnitude of the world to be modeled, aggregation may still be necessary for sake of computational reduction. The ecological foot-printing method for example is based on aggregate data, which could be coupled back to environment degradation, the notification of which may in return induce stronger emotions in agents to act, as is suggested by statistical data on urban and rural populations.

The above considerations indicate that the design and analysis of policies is a complex task. Many, highly interconnected, and unforeseen factors influence the applicability and result of new policies. This makes it hard to evaluate a (new) policy and foresee its implications. Macro-economic models are often fairly simple and founded on the principle of appropriate risk weighted return. System dynamics approaches [25] are often used for this effect and focus on the understanding of overall behavior of complex systems over time (e.g., causal feedback loops, nonlinearity). However, they are not well suited to study the effects on individuals and groups involved. That is, those models do not provide the instruments to evaluate policy at micro-level of implementation and are not able to handle uncertain situations. On the other hand, micro-models of individuals and groups, usually based on agent models for emergent global behavior, such as ABSS, do not provide the means to specify and regulate normative and regulative global restrictions [20,24].

Policy makers recognize the need for models that are able to incorporate different levels of abstraction and analysis. For instance, analysis of changes in criminal laws require both the understanding of its political and economic consequences at the macro level, but also the consequences on the working processes of criminal system staff, and the effects on the behavior of different risk groups. The above requirements indicate a need to incorporate macro-, and micro-level models in one simulation. We therefore propose a mediating layer, we call meso-layer, that connects the macro-, and micro-level in a simulation effective way. Basically in the meso layer we specify elements that "influence" the behavior of groups of agents. These are things like norms and organizational (or group) structure. These elements do not dictate behavior as elements on the micro level do, but are also different from the descriptive laws used on the macro level.

Section 2 describes our way to connect the macro- and micro-levels using this meso level. Section 3 motivates the need for extending existing agent models with personality, a hierarchy of needs, and a cultural and normative component. It also contains a short overview of existing work. The BRIDGE agent architecture presented in section 4 introduces a high level design for such agents. Ideas for future research and a summary of the proposals introduced in this paper are formulated in section 5.

2 Simulation Support for Policy Making

Every simulation model focusses on the aspects under study and abstracts from aspects that do not (or are not assumed) to significantly influence the aspects under study. Unintended effects are difficult to predict because analysis is mostly based on macroeconomic, econometric, models for policy analysis that are not able to represent individual interpretations, and thus cannot provide a clear linkage between structural features of the policy and the individual responses to it. E.g., a tax increase policy, expected to result in an increase of state revenue according to econometric models, may in fact have the overall effect of reducing tax revenue due to individuals' decisions not to earn money that is taxed or to opt for capital flight.

In complex systems, global behavior emerges unpredictably from the complex of individual micro-level behaviors of the autonomous and heterogenous actors, which are in turn influenced by the macro-level, that is, their own perception of the policies, their context, and their perception of the global outcome. Furthermore, the complex and dynamic nature of social phenomena increases due to the inter-actor relationships between the participating actors, which include individuals, groups, organizations and institutions. We claim that any useful simulation tool for policy making should be able to handle current problems faced by policy makers such as:

- representation of dynamic situations
- representation of individual behavior rather than averaged pseudo-rational behavior
- representation of inter-actor interaction
- representation of normative and cultural aspects
- visualization of above aspects and their dependencies

Our proposal for the next generation of simulation models is to create models that provide three layers of description (macro, micro and meso) and the connections between those layers. The most abstract, or macro, layer includes global functionalities and requirements. Macro-level expectations are specified as to enable a natural representation of the overall system that abstracts from (technical) issues. The lower, or micro, layer specifies the characteristics, aims and requirements of the individual entities and/or social groups. The middle layer allows for the coordination between micro and macro and enables the description and measurement of dynamic changes in the context. We will describe it further in the next subsection.

2.1 The Meso-level

From the macro level perspective, the meso level offers three types of components. If the macro level has a descriptive component that is empirically valid and which is not in the focus of the simulation (and thus not further expanded in the micro level) then this component is part of the meso level as a kind of "law of nature". I.e. it is assumed all agents (either individually or collectively) abide by this law. If the macro level contains a component like above, but which is in the focus of the simulation, then the component is merely treated as a benchmark to which agent behavior is compared. I.e. the micro level models of the agents should be such that this law "appears" in standard situations, but it does not have to appear when changes are made in the context (e.g. adding a new policy).

Finally, the meso level contains *new* components that are usually specified at society level in order to influence individual behavior. These components are things like norms, regulations and organizational structures but can also contain cultural biases. These components can be taken as given within the simulation framework. I.e. they do not arise from the agent behavior, but rather influence it top-down. E.g. a norm as *"it is forbidden to drive faster than 100 km/hr"* influences the decision of the agents how fast to drive. However, it depends on the individual agent whether it complies to the norm or not. This decision is in itself influenced by other norms (e.g. being on time at work), culture (e.g. we should all abide by the law) and personality (e.g. if the risk of being caught is low I don't follow the norm if it conflicts with my desires).

This third type of components is really new to the meso level. The other two types actually define aggregation and refinements within the simulation. Together the three types really make for a flexible but simulation-efficient framework.

The overall framework, depicted in figure 1 enables the definition of macro-level monitoring instruments and regulations, and the detailed specification of behaviors of individuals and cohese groups and their requirements. Furthermore, it enables the monitoring, specification and organization of committed and/or expected collective behavior, as basic concepts that complement existing macro and agent-modeling resources. The representation must include normative, ontological, organizational and evaluation

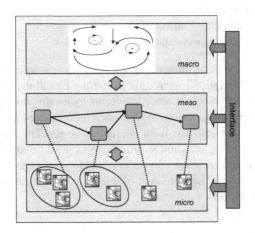

Fig. 1. Policy Making Framework

aspects. Whereas testing the effect of different rules (behavioral assumptions) on the system behaviors turbulence would be one thing, a next challenge would be to test the efficacy of policy strategies. The ideas on policy testing in agent based modeling as expressed in [13,14] form one starting point for this. The other is the work already done on the connection between meso-level elements such as norms and organizational structure to agents as is done in [28].

2.2 Interface Tools

Policy makers require methods and tools for support of policy design to be able to assess factors such as realization time, development cost, side effects, non-effects, negative effects and resistance to change. Furthermore, tools and techniques should enable participation in the development and analysis to a heterogeneous group of stakeholders, such as, politicians, regulators and the general public.

The policies imposed at the macro-, or mesa-level will be reasoned about at the agent-level, such that the agents are capable making their decisions with respect to the proposed/imposed policies and the mechanisms that are put in play to enforce compliance with the policies. This requires of course that agents need to be equipped with a cognitive representation of their social environment in order to capture the social dynamics in their society. Furthermore, the policies should address the behavioral drives and processes of the individual agents.

In order to analyze and understand the consequences of policies, complexity must be captured at all three levels. It is therefore essential to ensure that the models are easy to use and give interesting, understandable projections that usefully inform scenario development in an engaging and productive framework for stakeholder participation. Such a framework requires interface tools that visualize the dynamics and statistics of the simulation in comprehensive formats and that enable the users to change assumptions and settings quickly and easily, so that policy makers can experiment by running various simulations. Due to space limitations, the focus of this paper is on the underlying model for policy design and not on the interface.

3 Personality, Culture, Norms, Hierarchy of Needs

Representations of human behavior in social simulations entail models for deciding about agents intentions, based on agents beliefs and desires. According to March [21], decision making processes may be rational or rule following. Rational decision making aims to maximize a utility function and is used in economic games and simulations. Human decision making processes often have both rational and rule-following aspects. Rule-following decision making can be seen as imposing moral boundaries on acceptable outcomes of rational decision making. It can also be seen as consolidated experience or an evolutionary outcome of rational decision making [21]. In our opinion, agents in simulations for the purpose of policy making should decide in a more human fashion and apply both types of decision making. This implies that agents must grasp the concept of norms, have cultural scripts, possess a personality, and react based on a hierarchy of needs in correspondence to human nature. In the rest of this paper we will describe how agents can be endowed with these elements, which are introduced below.

3.1 Hierarchy of Needs

Individual behavior follows from basic motivations: the satisfaction of natural needs (hunger, etc.) the difference laying on the way these are satisfied [2]. These motivations (also called source of actions by) make an agent behave either reactively or cognitively (i.e. adopting a low or a high level behavior). Maslow's Pyramid of Needs synthesizes a large body of research related to human motivation [22] that views all needs as instinctive, but some are more powerful than others. Needs in the lower levels are more powerful, while the highest level is only acted upon when the other levels are met. The bottom levels (Physiological, Security, Social and Esteem Needs) of the pyramid are called "deficiency needs": the individual does not feel anything if they are met, but feels anxious if they are not met. Failure to meet these needs leads to deprivation. The highest-level of the pyramid is called a growth need: when fulfilled, they do not go away; rather, they motivate further. Growth needs do not stem from a lack of something, but rather from a desire to grow as a person.

3.2   Norms

Norms can be seen as constraints imposed by society on the behavior of the individuals. An important aspect of norms is that individuals can decide to violate them. The violation can itself be a trigger of further behavior in order to punish the violation. Thus norms do not directly constrain behavior but influence the decisions to take a certain course of action. They can do this in different ways. A person may be aware of a norm like "one should always give an answer to a request" even though he does not agree with it. He can exploit the existence of the norm by always first requesting information before trying to find it himself. He does not necessarily comply to the norm and if he does it might be an unconscious decision or because it benefits him.

A person can also accept a norm. In that case he agrees that the norm is a good one and he tries to follow it as much as possible. Only in special situations (e.g., if the norm contradicts another, more important norm) will he violate it.

We assume that norms are typically explicitly available, allowing the agent to reason with and about them. Reasoning explicitly about norms is included in a number of social simulations, see, e.g., [8,5].

3.3   Personality

Personality has been defined as *"the distinctive and characteristic patterns of thought, emotion, and behavior that define an individual's style and influence on his or her interactions with the environment"* [9]. Personality is not something concrete, and existing personality measurements or indicators can just give us an indication of one's personality. A personality inventory commonly used is the Myers-Briggs Type Indicator (MBTI), based upon Carl Jung's theories [16,18]. According to Jung, people take in and process information in different ways. He characterized individuals in terms of attitudes (extraversion and introversion), perceptual functions (intuition and sensing) and judgement functions (thinking and feeling). Meyers-Briggs extended this classification with a forth dimension that distinguishes between judgement and perception:

Extraversion or Introversion: Shows how a person orients and receives their energy. Extravert people prefer to acquire their personal energy from the outer world of people and activities, as where introverts prefer to acquire their personal energy from the inner world of ideas and thoughts.

Sensing or Intuition: Describes how people take in information. Sensing people tend to concentrate primarily on the information gained by the senses, creating meaning from conscious thought and limiting their attention to facts and solid data. People who prefer intuition are interested in the "big picture" when interpreting information, taking a high-level view, as opposed to digging into the detail. They concentrate on patterns, connections and possible meanings.

Thinking or Feeling: Indicates how people prefer to make decisions. People who prefer thinking base their decisions on logic, objective analyzes of cause and effect, whereas people who prefer feeling are influenced by their concerns for themselves and others.

Judging or Perceiving: Describes the way you manage your life and how you deal with the outer world. People who prefer judging like to have a planned lifestyle, everything should be in order and in a scheduled manner. On the contrary, people who prefer perceiving are more unplanned, flexible and spontaneous in their lifestyle. They prefer to keep all options open.

Personality determines how decisions are taken and basic patterns of behavior (and therefore will determine the type of reasoning used by simulation agents). E.g. a sensing person needs to have all facts established before taking a decision while an intuitive person concentrates on possibilities. The explicit representation of personality as we propose allows the agent introspection on its own personality; a feature compared to most ABSS frameworks.

3.4 Culture

In all aspects of human life, the desires of people, the decisions people make and the procedures for decision making are culture-dependent in several ways. First, the priority of goals depends on culture; for instance "maximize personal wealth" may have priority over "maintain pleasant interpersonal relation". Second, preferences for rational versus rule following procedures differ across cultures. Third, if a rule following procedure is chosen, the rules depend on culture. Fourth, a decision may be interpreted offensive by an opponent having a different cultural background. Also, the appropriate reaction to inappropriate behavior differs across cultures.

Culture is what distinguishes one group of people from another [11]. This implies that culture is not an attribute of individual people, unlike personality characteristics. It is an attribute of a group that manifests itself through the behaviors of its members. For a trading situation, as analyzed in [12], the culture of the trader will manifest itself in four ways. First, culture filters observation. It determines the salience of clues about the acceptability of trade partners and their proposals. Second, culture sets norms for what constitutes an appropriate partner or offer. Third, it sets expectations for the context of the transactions, e.g., the enforceability of regulations and the possible sanctions in case of breach of the rules. Fourth, it sets norms for the kind of action that is appropriate given the other three, and in particular, the difference between the actual situation and the desired situation. Of course, culture is also created by people and thus generated by the behavior they display. However, we assume that the time scale on which culture changes is much larger than what we are typically interested in during a simulation period. Therefore we assume the culture to be a fixed component rather than an evolving entity itself (for our purpose).

4   BRIDGE Agent Architecture

In order to develop agents that are able to reason about complex social situations, such as determine the consequences of policies to themselves and decide to adapt their behavior accordingly, it is essential to provide agents with constructs for social awareness, 'own' awareness and reasoning update. The fact that these issues are explicitly represented in the agent architecture allows for introspection on the drives of behavior.

Existing agent models can be divided into two types: deliberate reasoning models and unreasoned models[1]. Both have their strengths and weaknesses. In fact, Silverman states *"no model will ever capture all the nuances of human emotion, the full range of stress effects, or how these factors affect judgment and decision making. However, to the extent that a model provides a valid representation of human behavior, it will likely be useful to those who wish to simulate that behavior"* [24].

Deliberate agent models, such as BDI [3], have formal logic-grounded semantics, but require extensive computational resources. The Subsumption Architecture [4] takes an intentional view, and aims to provide the behavior displayed by lower level life forms which Brook's claims are *"almost characterizable as deterministic machines"*, through a combination of simple machines with no central control, no shared representation, slow switching rates and low bandwidth communication. Cognitive models, such as ACT-R and SOAR [1] aim at understanding how people organize knowledge and produce intelligent behavior based on numerous facts derived from psychology experiments, and employing quantitative measures. However, these models lack realism since they do not incorporate demographics, personality differences, cognitive style, situational and emotive variables, group dynamics and culture. On the other hand, neurological oriented models that mimic the brain, such as neural networks, lack transparency to link observed behavior to the implementation. Realistic agent models should combine the characteristics of the different types.

We are interested in understanding the consequences of macro-level design (policies) on the micro-level decisions of individual agents. This is in some sense the opposite of [24] where the aim is the *"emergent macro-behavior due to micro-decisions of bounded rational agents"*. The model of the human mind CLARION [27] aims to explore the interaction of implicit and explicit cognition, emphasizing bottom-up learning (i.e., learning that involves acquiring first implicit knowledge and then acquiring explicit knowledge on its basis). CLARION's goal is to form a (generic) cognitive architecture that captures a variety of cognitive processes in a unified way and thus to provide unified explanations of a wide range of data. The CLARION model and the BDI model are both excellent candidates for the extension as aimed for in this paper. The emphasis on deliberation in the study of the reaction to new policies, gives us a

---

[1] We use this word to mean the opposite of deliberate decision making. We refer to such notions as automatic, inborn, inherent, innate, instinctive, intuitive, involuntary, native, mechanical, natural, reflex, spontaneous, unlearned, unpremeditated, unthinking, and visceral. This is thus different from the more usual concept of 'reactive'.

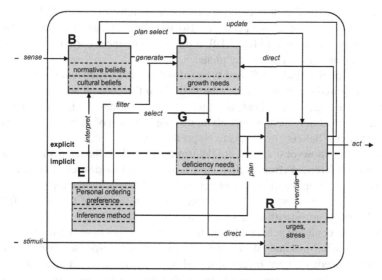

Fig. 2. The BRIDGE Agent Architecture

slight preference for the BDI model over the CLARION model. However, the ideas presented here can also be applied to the CLARION model.

Our aim is to describe, given a new policy, how different people react to it when made aware of the new policy. The decisions made by the agent in reaction to the new policy are the basis for the adaptation of the agents' behavior, which in its turn will determine the emergence of changed macro-behavior, the consequence of the policy. In this view, policy analysis at the agent level is a rational activity, for which agents explicitly reason based on their beliefs, desires and intentions. In this paper we add to this that this decision making process is not only deliberate, but also unreasoned. As a consequence of the above, policy-analysis agents require the extension of the BDI representation to support a description of the agent's personality, hierarchy of needs, and normative-cultural context. For this reason, we propose the BRIDGE agent architecture (Beliefs, Response, Intentions, Desires, Goals and Ego) as depicted in figure 2.

4.1 BRIDGE Mental Components

The mental components of the BRIDGE agent architecture incorporate the explicit rational process and implicit (unreasoned) behavioral aspects. BRIDGE is so defined that it can be used in various settings: 'pure' emergent behavior is achieved by using only the subsystems in the lower level of the architecture, purely intentional deliberate behavior results from the top level of the architecture, while the combination of both levels integrate deliberate and unreasoned decision making.

Ego. Describes the filters and ordering preferences that the agent uses. Personality type determines the choice of reasoning, e.g., backward reasoning (explorative, goal-based) is associated with intuitive types while forward (belief- or evidence-based) reasoning is typical of Sensing types.

Response. Relates to the bottom layers of the hierarchy of needs and implements the reactive behavior of the agent. It also includes representation of fatigue and stress coping mechanisms [10,15,19]. It directly influences current goals and when needed can and will overrule any plans.

Beliefs. The model includes different types of beliefs and beliefs over beliefs. Beliefs can be acquired through senses, 'inherited' from the cultural and normative background of the agent, or indicate the agent's own judgement on its context. Furthermore, the agent may have beliefs about its own ordering of its beliefs (e.g., believe that normative beliefs are always to be preferred).

Desires. Besides those desires that are determined based on the agents beliefs and its personal preferences (from Beliefs and Ego), the set of desires also includes the growth needs from the hierarchy of needs. Growth needs are the top of the Maslow pyramid, and can be seen as maintenance goals in the sense that when 'fulfilled, they do not go away; rather, they motivate further'.

Goals. Possible goals are generated from the current Desires based on the agent's preferences, possibilities and current state (from Ego and Response) together with the deficiency needs (bottom layers of the Maslow pyramid) which are always in the set of goals. The intuition behind this, is that deficiency needs are always a goal, but not an intentional one in the sense that 'the individual does not feel anything if they are met, but feels anxious if they are not met'. Response factors such as fatigue or stress direct the choice and priority of goals.

Intentions. Contains possible plans to realize goals, based on agent capabilities and preferences. However, when Response determines a high level of need, it will overrule any explicitly formed goals with lower level needs.

4.2   BRIDGE Reasoning Process

The BRIDGE reasoning process is carried by the components just described and the interactions between them, see 2. The components and the interactions between them have to work concurrently to allow continuous processing of the input in the form of sensory information and of effects on the body and brain as modeled by the arrow "stimuli". All these links are discussed below.

sense. Consciously received input from the environment, such as messages and observations. In the Beliefs subsystem, sensory input is interpreted in the form of belief updates of the agent.

stimuli. Unconsciously received influences. In the Response subsystem, stimuli are directly processed resulting in an emotional state that prepares for action. In this way we can model unreasoned (involuntary) responses. Note that the actions carried out can have effects that can sensed by the agent, conform [6].

interpret. Personality characteristics provide different interpretations of beliefs, by adding extra weight or priority to some beliefs.

filter. Personality (ego) characteristics indicate possible instantiation and ordering of beliefs. This functions as a filter on the desires of the agent at any moment. Different personalities give different priorities to certain types of goals, e.g., based on cultural background (such as the desire to accumulate and show one's wealth) and determine different ways the agent will interact in its social setting [23,7].

select. Possible goals are selected from current desires. Based on personality charac-teristics (Ego) choice and ordering of goals and deficiency needs is determined.

plan. The calculation of possible plans is influenced by Ego and Growth.

generate Desires are generated from beliefs. This includes desires that are based on the agent's perception of its normative and cultural background.

plan select. The set of current beliefs has an impact on the plan selection process per-formed by Intentions.

update. Beliefs are possible updated.

act Plans are performed in the environment through this link.

direct. Basic urges, the current emotional state, and stress levels direct the order and choice of current goals.

overrule. Basic urges can become so strong that they overrule any (rationally formed) plans and immediate action is taken corresponding to the basic urge.

5   Conclusions

In this paper we have set out a preliminary framework for building simulation support for policy making. We argued that such a simulation tool needs to combine both macro- and micro-level models. In order to combine these two levels we introduced a meso-level, which serves on the hand as a place to aggregate some micro-level data for the macro-level models as well as give macro-level input for the micro-level agents. How-ever, we claim that an at least equally important function of the meso-level is to specify society level elements, such as norms, that influence the agents on the micro-level. These elements do not emerge from the micro-level neither are they used as laws of nature from the macro-level. Rather they are used as inputs for the deliberation process of the agents.

In order for the agents to realistically simulate human behavior, we argue that they should also incorporate mechanisms to cope with these meso-level elements. Our pro-posed BRIDGE model provides for potentially very rich agents. They might also be very computationally inefficient if every action would require deliberation over all these components. However, we also provide a "unreasoned", reactive layer that can be used to bypass the deliberation. This provides the opportunity to use the deliberation only for those elements that are of prime importance for the simulation while having reac-tive behaviour for standard situations.

Although a new framework is easy to specify, the question is how this framework should be implemented. In our case we envision the agents to be implemented in an ex-tension of 2APL which can be connected to an existing simulation tool such as Repast to provide a proper simulation environment. Finally, we will use the the OperA framework to define the meso-level elements, and generate patterns to be used by the simulation tool at the micro-level. Thus we have a basis for starting to actually build the simulation tool for policy making.

Acknowledgements. The authors are grateful to W. Jager and B. Hoevenaars and the anonymous reviewers for their contributions and comments to earlier versions of this paper. The research of V. Dignum is funded by the Netherlands Organization for Scien-tific Research (NWO), through Veni-grant 639.021.509.

References

1. Anderson, J.R.: The adaptive character of thought. Lawrence Erlbaum, Mahwah (1990)
2. Andriamasinoro, F., Courdier, R.: Integration of generic motivations in social hybrid agents. In: Lindemann, G., et al. (eds.) RASTA 2002. LNCS, vol. 2934, pp. 281–300. Springer, Heidelberg (2004)
3. Bratman, M.: Intention, Plans, and Practical Reason. CSLI Publications (1987)
4. Brooks, R.A.: How to build complete creatures rather than isolated cognitive simulators. In: VanLehn, K. (ed.) Architectures for Intelligence, pp. 225–239. Lawrence Erlbaum, Mahwah (1991)
5. Castelfranchi, C., Dignum, F., Jonker, C.M., Treur, J.: Deliberative normative agents: Principles and architecture. In: Jennings, N.R. (ed.) ATAL 1999. LNCS (LNAI), vol. 1757, pp. 364–378. Springer, Heidelberg (2000)
6. Damasio, A.: The Feeling of What Happens: Body, Emotion and the Making of Consciousness. Harcourt Brace (1999)
7. Dastani, M., Dignum, V., Dignum, F.: Role assignment in open agent societies. In: AAMAS 2003. ACM Press, New York (2003)
8. Dignum, F., Edmonds, B., Sonenberg, L.: The use of logic in agent-based social simulation. JASSS 7(4) (2004)
9. Eladhari, M., Lindley, C.: Player character design facilitating emotional depth in mmorpg's. In: Digigal Games Research Conference (2003)
10. Folkman, S., Lazarus, R.S.: Coping and emotion. In: Stein, N.L., Leventhal, B., Trabasso, T. (eds.) Psychological and Biological Approaches to Emotion, pp. 313–332. Lawrence Erlbaum, Mahwah (1990)
11. Hofstede, G.: Cultures Consequences, 2nd edn. Sage Publications, Thousand Oaks (2001)
12. Hofstede, G.J., Jonker, C.M., Meijer, S., Verwaart, T.: Modelling trade and trust across cultures. In: Stølen, K., Winsborough, W.H., Martinelli, F., Massacci, F. (eds.) iTrust 2006. LNCS, vol. 3986, pp. 120–134. Springer, Heidelberg (2006)
13. Jager, W.: The four p's in social simulation, a perspective on how marketing could benefit from the use of social simulation. Journal of Business Research 60(8), 868–875 (2007)
14. Jager, W., Mosler, H.J.: Simulating human behavior for understanding and managing environmental dilemmas. Journal of Social Issues 63(1), 97–116 (2007)
15. Janis, I.L., Mann, L.: Decision making: A psychological analysis of conflict, choice, and commitment. The Free Press (1977)
16. Jung, C.: Psychologische Typen. Rascher Verlag (1921)
17. Kable, J.W., Glimcher, P.W.: The neural correlates of subjective value during intertemporal choice. Nat. Neuroscience 10(12), 1625–1633 (2007)
18. Keirsey, D.: Please understand me II. Temperament Character Intelligence. Prometheus Nemesis Book Company (1998)
19. Lazarus, R.S., Folkman, S.: Stress, Appraisal, and Coping. Springer, Heidelberg (1984)
20. Macy, M.W., Willer, R.: From factors to actors. Annual Review of Sociology 28 (2002)
21. March, J.G.: A Primer on Decision Making: How Decisions Happen. Free Press (1994)
22. Maslow, A.: Motivation and Personality. Harper (1954)
23. Sichman, J., Conte, R.: On personal and role mental attitude: A preliminary dependency-based analysis. In: de Oliveira, F.M. (ed.) SBIA 1998. LNCS, vol. 1515, pp. 1–10. Springer, Heidelberg (1998)

24. Silverman, B., Johns, M., Cornwell, J., O'Brien, K.: Human behavior models for agents in simulators and games: Part i: Enabling science with pmfserv. Presence: Teleoperators and Virtual Environments 15(2)

25. Sterman, J.: Business Dynamics System Thinking and Modeling for a Complex World. McGraw-Hill Higher Education, New York (2000)

26. Stern, N.: Stern review report on the economics of climate change. HM Treasury, Independent Reviews (2006)

27. Sun, R.: The clarion cognitive architecture: Extending cognitive modeling to social simulation. In: Cognition and Multi-Agent Interaction. Cambridge University Press, Cambridge (2006)

28. Vazquez-Salceda, J., Dignum, V., Dignum, F.: Organizing multiagent systems. JAA-MAS 11(3), 307–360 (2005)

A Quantitative Method for Comparing Multi-Agent-Based Simulations in Feature Space

Ryota Arai and Shigeyoshi Watanabe

The University of Electro-Communications
1-5-1 Chofugaoka, Chofu, Tokyo 182-8585, Japan
{ryota,watanabe}@ice.uec.ac.jp

Abstract. Comparisons of simulation results (model-to-model approach) are important for examining the validity of simulation models. One of the factors preventing the widespread application of this approach is the lack of methods for comparing multi-agent-based simulation results. In order to expand the application area of the model-to-model approach, this paper introduces a quantitative method for comparing multi-agent-based simulation models that have the following properties: (1) time series data is regarded as a simulation result and (2) simulation results are different each time the model is used due to the effect of randomness, even though the parameter setups are all the same. To evaluate the effectiveness of the proposed method, we used it for the comparison of artificial stock market simulations using two different learning algorithms. We concluded that our method is useful for (1) investigating the difference in the trends of simulation results obtained from models using different learning algorithms; and (2) identifying reliable simulation results that are minimally influenced by the learning algorithms used.

Keywords: validation, multi-agent-based simulation, time series classification, model-to-model approach, reinforcement learning.

1   Introduction

The validation of computational models and simulation results is a critical issue in agent-based simulation because the simulation results are very sensitive to the manner in which agents are modeled [1]. By considering this issue, several approaches have been proposed for the validation of social simulations. Axtell et al. [2] proposed the concept of "alignment of computational models" or "docking" that can be used to examine whether two computational models can produce the same results. Both the computational models are minimally validated if their results are identical. Takadama et al. [3], however, indicated that the following difficulties arose from the use of this approach: (1) it is difficult to compare different computational models fairly by using the same evaluation criteria since the models are developed for different purposes; (2) there are very few common elements in different computational models, which makes it difficult to replace either computational model with the other; and (3) simulation results are sensitive to even small modifications in a model, which complicates the identification

N. David and J.S. Sichmann (Eds.): MABS 2008, LNAI 5269, pp. 154–166, 2009.
© Springer-Verlag Berlin Heidelberg 2009

of the key elements or factors in a model that are responsible for the sensitivity of the simulation results. In order to overcome these difficulties, they proposed a cross-element validation method that validates computational models by investigating whether different models can produce the same results when an element in the agent architecture is changed. This method is based not on the between models but on a within model. The concept and method mentioned above can be regarded as involving model-to-model approaches.

One of factors preventing the widespread application of these approaches is the lack of methods for comparing multi-agent-based simulation results. Earlier studies focused on models whose simulation results were single numerical values. In the case of multi-agent-based simulation models, especially used in artificial society, however, a variety of artificial market models have been studied. These models are found to have the following properties: (1) time series data is regarded as a simulation result; and (2) simulation results are different each time the model is used due to the effect of randomness, even though the parameter setups are identical in every use. In order to expand the application area of the model-to-model approach, it is necessary to establish a comparison method for multi-agent-based simulation models. In this paper, we propose a quantitative comparison method for these models and demonstrate its use by using it for the comparison of artificial stock market simulations.

This paper is organized as follows. Section 2 presents the proposed comparison method. The artificial stock market model and the learning algorithms used are described in Section 3. Section 4 presents the results of an evaluation of our proposed method. Section 5 discusses related research studies, and Section 6 presents the conclusion.

2   Quantitative Comparison Method

Simulation results involving single numerical values can be compared in the coordinate system. However, such comparison is not easy when the simulation results are in the form of time series data, similar to those obtained in stock market simulation. It is also not meaningful to compare the simulation results obtained by using different models only once because the results of every simulation are different due to the effect of randomness, even though the parameter setups are identical in every use. In order to overcome these difficulties, we developed a method that collects multiple simulation results from models with the same parameter setup, these results were obtained using different random seeds. Subsequently, the distance between two distributions of the multiple simulation results was calculated to evaluate the difference in the models quantitatively. Our method reduces the high dimension of time series data by employing a feature extraction technique. Figure 1 shows an overview of the proposed comparison method; a simplified algorithm of the method is described as follows:

1. Multiple time series simulations of models are performed under different random seeds w_i and w_i' using the same parameter setup A and setup B, respectively, where $i = 1, 2, \cdots, s$ and s is the number of simulation results,

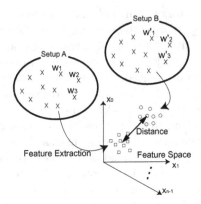

Fig. 1. Overview of the proposed comparison method; each point w_i or w'_i in the distributions represents time series data obtained as a simulation result

2. Features of each time series simulation result are extracted and mapped to an n-dimensional feature space, where n is the number of time series data,
3. The distance between two distributions in the feature space is measured.

We discuss the feature extraction technique in the following section, and the measurement of the distance between two distributions is dealt with in section 2.2. In section 2.3, we introduce a visualization technique to demonstrate the similarities and dissimilarities relations between different distributions.

2.1   Feature Extraction from Time Series Data

The classification of time series data has been an active research topic in the field of data mining fields [4]. The high dimensionality of time series data presents a major problem in their classification. The high dimensionality of data gives rise to the "curse of dimensionality" and problems related to distance metrics [5]. For example, in high-dimensional space, the distance between a given point and the nearest data point approaches the distance between the given point and the farthest data point [6]. However, all the data point need not be used because time series data exhibit considerable redundancy. Thus, feature extraction should be used to reduce the dimensionality of time series and to retain only the important information.

When a feature extraction technique is used to compress time series data, it is important to preserve the maximum possible amount of energy of the original data. In the case of market simulations, most of the financial time series data can be regarded as pink noise or 1/f noise (which is a signal with a frequency spectrum in which the power spectral density is proportional to the reciprocal of the frequency. This means that powers of the time series data is massing in the low frequency spectrum). On the basis of this assumption, we perform feature extraction by using the discrete Fourier transform (DFT). This feature extraction technique was proposed by Agrawal et al. [7]. The DFT maps the time series data from the time domain to the frequency domain. The distance between arbitrary two points in the time domain is scale-invariant with the distance between the mapped two points in the frequency domain.

The feature extraction process is shown as follows:

1. x_t is calculated by multiplying time series data and the humming window $w_t = 0.54 + 0.46 \cos\left(\frac{2\pi}{n}t - \pi\right)$ at time t, where $t = 0, 1, \cdots, n-1$,
2. The sequence $\vec{x} = [x_t]$, in the time domain is transformed into a sequence of n complex numbers $\vec{X} = [X_f]$, where $f = 0, 1, \cdots, n-1$, in the frequency domain by using the following formula:

$$X_f = \frac{1}{\sqrt{n}} \sum_{t=0}^{n-1} x_t \exp\left(-j\frac{2\pi f t}{n}\right) \tag{1}$$

where j is the imaginary unit.

In the feature extraction technique, after the transformation of the time series data from the time domain to the frequency domain, only the first few coefficients are retained as features and all other coefficients are discarded to reduce the dimensionality of the data.

2.2   Distance Measure

We introduce a distance measure to evaluate the difference between two distributions of simulation results. By employing the feature extraction technique, each simulation result is mapped to an n-dimensional feature space. Thus, the distance between two distributions of simulation results is regarded as the distance between two distributions of features in the feature space, as shown in figure 1. To measure the distance between distributions, we employ the Mahalanobis generalized distance [8], in that it takes into account not only the mean values of the data set but also the correlations of the data set and is scale-invariant. The Mahalanobis generalized distance is defined as

$$D_M^2(m_1, m_2) = (m_1 - m_2)^t \Sigma_W^{-1}(m_1 - m_2) \tag{2}$$

where m_1 and m_2 are mean vectors of two distributions and Σ_W refers to the within-class covariance matrix that is defined as

$$\Sigma_W = \sum_{i=1,2} P(\omega_i)\Sigma_i \tag{3}$$

$$= \sum_{i=1,2}\left(P(\omega_i)\frac{1}{n_i}\sum_{x\in\mathcal{X}_i}(x - m_i)(x - m_i)^t\right) \tag{4}$$

where $P(\omega_i)$ and x are a priori probability of and feature vector in class ω_i, respectively. If the number of samples (simulation results) in classes ω_1 and ω_2 are identical, then the relation $P(\omega_1) = P(\omega_2) = \frac{1}{2}$ holds.

Our comparison method can evaluate the difference between only two distributions. When multiple distributions are to be compared, we measure the distance between every two distributions and construct a distance matrix such as that shown below.

$$M_d = \begin{array}{c} \\ 1 \\ 2 \\ \vdots \\ n \end{array} \begin{array}{cccc} 1 & 2 & \cdots & n \\ \left(\begin{array}{cccc} 0 & d_{12} & \cdots & d_{1n} \\ d_{21} & 0 & \cdots & d_{2n} \\ \vdots & & \ddots & \vdots \\ d_{n1} & d_{n2} & \cdots & 0 \end{array} \right) \end{array} \tag{5}$$

Here, d_{ij} represents the distance between distributions i and j; further, $d_{ij} = d_{ji}$ and $d_{ii} = 0$.

2.3   Visualization Technique

In order to analyze the results of the comparison of models and to examine the validation of models, it is important to visualize the similarities and dissimilarities among multiple distributions. Generally, the distance between two points $P_i(x_i, y_i)$ and $P_j(x_j, y_j)$ is defined as

$$d_{ij} = \sqrt{(x_i - x_j)^2 + (y_i - y_j)^2} \tag{6}$$

The distance can be computed uniquely when the coordinates of the two points are given. However, the coordinates of two points cannot be determined uniquely when only the distance between the points is known. Multidimensional scaling (MDS) is one of the solution methods for obtaining the coordinates of points from the distance between the points [9]. From the result of MDS, the points can be visualized in a geometric space of low dimensionality (usually, two-dimensional space). Using this visualization technique, the similarities and dissimilarities between different distributions are determined.

Let us assume that the coordinates of two objects $\mathbf{x_i}$ and $\mathbf{x_j}$ are given as $(x_{i1}, x_{i2}, \cdots, x_{ip})$ and $(x_{j1}, x_{j2}, \cdots, x_{jp})$, where p is the dimensionality of the space. Let k denote the origin of the space $(0, 0, \cdots, 0)$. Then, the inner product z_{ij} of two objects $\mathbf{x_i}$ and $\mathbf{x_j}$ is as follows:

$$z_{ij} = d_{ik} d_{jk} \cos \theta = \sum_{m=1}^{p} x_{im} x_{jm} \tag{7}$$

For the triangle ijk, the following equation holds:

$$d_{ij}^2 = d_{ik}^2 + d_{jk}^2 - 2 d_{ik} d_{jk} \cos \theta \tag{8}$$

Therefore, the inner product can be represented as

$$z_{ij} = \frac{1}{2} \left(d_{ik}^2 + d_{jk}^2 - d_{ij}^2 \right) \tag{9}$$

Table 1. Condition Bits

Bit	Condition
1	Price * interest/dividend > 1/4
2	Price * interest/dividend > 1/2
3	Price * interest/dividend > 3/4
4	Price * interest/dividend > 7/8
5	Price * interest/dividend > 1
6	Price * interest/dividend > 9/8
7	Price > 5-period MA
8	Price > 10-period MA
9	Price > 100-period MA
10	Price > 500-period MA

Since z_{ij} is given as $\sum_{m=1}^{p} x_{im}x_{jm}$, the inner product matrix for z_{ij} is

$$\mathbf{Z} = \mathbf{X}\mathbf{X}^t \tag{10}$$

where $\mathbf{X}^t$ denotes the transpose of $\mathbf{X}$. In order to obtain $\mathbf{X}$, we consider the minimization of an objective function Q that is defined as

$$Q = \sum_i \sum_j \left(z_{ij} - \sum_{m=1}^{p} x_{im}x_{jm} \right)^2 \tag{11}$$

To minimize Q, we use the Eckart and Young [10] decomposition of a matrix. First, we calculate the eigenvalues (denoted by $\lambda_1, \cdots, \lambda_p$) and eigenvectors (denoted by $\mathbf{y}1, \cdots, \mathbf{y}^\mathbf{p}$) of $\mathbf{Z}$. Subsequently, by using a diagonal matrix of eigenvalues and a matrix of eigenvectors, denoted by $\mathbf{\Lambda}$ and $\mathbf{Y}$, respectively, we obtain the following equation:

$$\mathbf{Z} = \mathbf{Y}\mathbf{\Lambda}\mathbf{Y}^t \tag{12}$$

From this decomposition, we obtain $\mathbf{X} = \mathbf{Y}\mathbf{\Lambda}^{\frac{1}{2}}$.

3   The Model

3.1   An Artificial Stock Market Model

To evaluate the effectiveness of the proposed comparison method, we constructed an artificial stock market model based on the early Santa Fe artificial stock market model [11]. The market has only a single stock that is priced at $p(t)$ per share at time t, there are N agents labeled with $i = 1, 2, \cdots, N$. Time is discrete ($t = 1, 2, \cdots$); period m lasts from time t until time $t + 1$. The stock pays a dividend of $d(t + 1)$ per share at the end of period m. The dividend time series $d(t)$ is itself the result of a stochastic process; the process is defined to be independent of the market and the agents' actions. At any given time t, each agent i has a particular number of shares $h(t)$ and assesses the market conditions [12] on the basis of table 1. Each condition in the table can be either true or false; m-period MA denotes the moving average over the most recent m steps of $p(t)$, i.e.,

$$MA(t,m) = \sum_{i=0}^{m-1} \frac{p(t-i)}{m} \tag{13}$$

Subsequently, each agent submits either a bid to buy $b_i(t)$ shares or an offer to sell $o_i(t)$ shares on the basis of their own decision-making process. Here the total bid and total offer for the stock at time t are respectively given by

$$B(t) = \sum_{i=1}^{N} b_i(t) \tag{14}$$

$$O(t) = \sum_{i=1}^{N} o_i(t) \tag{15}$$

The number of shares $h_i(t)$ with an agent i is defined as

$$h_i(t) = h_i(t-1) + \frac{V(t)}{B(t)} b_i(t) - \frac{V(t)}{O(t)} o_i(t) \tag{16}$$

where $V(t) = min[B(t), O(t)]$ is the number of transactions. The stock price $p(t)$ depends on the overall buying and selling trends of the agents.

$$p(t+1) = p(t)\{1 + \eta(B(t) - O(t))\} \tag{17}$$

The parameter η is an adjustable parameter that influences the stock price. At the end of each period m, each agent obtains a reward that can be expressed as

$$reward = \pm (p(t) - p(t-1) + d(t) - r \cdot p(t-1)) \tag{18}$$

where r is the interest rate.

3.2   Modeling Agents

For the purpose of cross-element validation, we employ Q-learning and SARSA [1] to update the worths of pairs of states and action in our simulation. These learning systems are expressed in the form of eq.(19) and eq.(20). The variables in these equations are explained in Table 2.

$$\text{Q–learning} : Q(s,a) = Q(s,a) + \alpha[r + \gamma\max_{a' \in A(t')} Q(s',a') - Q(s,a)] \tag{19}$$

$$\text{SARSA} : Q(s,a) = Q(s,a) + \alpha[r + \gamma Q(s',a') - Q(s,a)] \tag{20}$$

Table 2. Variables in Q-learning and SARSA

$Q(s, a)$	the worth of selecting action a in state s
$Q(s', a')$	the worth of selecting the next action a' in the next state s'
r	the reward corresponding to the payoff obtained
$A(s')$	a set of possible actions in the next state s'
$\alpha(0 < \alpha \leq 1)$	learning rate
$\gamma(0 \leq \alpha \leq 1)$	discount rate

In this study, the actions were selected on the basis of the Boltzmann distribution selection method. In this method, an action is probabilistically selected on the basis of the ratio of Q-values for all actions; this ratio is calculated by the following equation

$$p(a|s) = \frac{e^{Q(s,a)/T}}{\sum_{a_i \in A} e^{Q(s,a_i)/T}} \qquad (21)$$

where T is the temperature that adjusts the randomness of action selection. Agents select their actions at random when T is high; when T is low, they select greedy actions.

4   Experiments on Artificial Stock Market

4.1   The Relationship between Distance and the Number of Simulation Results

The purpose of this experiment is to investigate the relationship between the distance obtained from eq. (2) and the number of simulation results in a distribution. We set the number of simulation results to be between 10 and 1000. Although each simulation was conducted in 5000 steps, only the last 4096 steps were considered and those from the first 904 steps were ignored; this was because the first 904 steps were considered to comprise the initial learning period. We then employed the feature extraction technique to reduce the data and retained the first four Fourier coefficients.

Figure 2 shows the result of the experiment. In the experiment, multiple parameter setups were used; however, only the results pertaining to four cases are shown in Figure 2. The other results also show the same trend. The figure indicates that stable distances are obtained by measuring the distance between distributions that include a specific number of simulation results. Thus, in the next experiment, we set the number of simulation results in a distribution to be 100. Although, this is expected to provide a more stable result due to the increase in the number of simulation results in a distribution, it results in an increase in the computational time, which is undesirable.

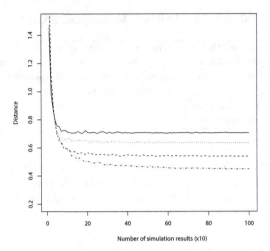

Fig. 2. The relationship between distance and the number of simulation results

4.2   Comparison of Q-Learning with SARSA

The purpose of this experiment is to investigate the difference between simulation results obtained by employing different learning algorithms. We believe that when similar results are derived from models using different learning algorithms, the results have a certain degree of validity. This is because the simulation results are minimally influenced by the manner of implementation of the learning algorithms; this in turn prevents arbitrary modeling. Through this experiment, we also show that our comparison method is effective in finding valid simulation results. In this experiment, we employed two types of models – one that used Q-learning and another that used SARSA. The learning rate α was set to be in the range 0.0 to 1.0 for each model. We assume that α influences the simulation result trends because the difference between Q-learning and SARSA arises from the second terms in eqs. (19) and (20). The number of agents was set to be 30 and the other parameters were the same in both cases. Figure 3 shows the result of the experiment. We located all the distributions in the simulation results in two-dimensional space by employing multidimensional scaling. QL(α) denotes the model that uses Q-learning and SA(α) denotes the model that uses SARSA.

From figure 3, the following information can be obtained

1. when $\alpha = 0.0$, the simulation results appear as anomalous points;
2. when $\alpha \neq 0.0$, the simulation results exhibit a different trend;
3. when α is small, the simulation results are similar for both types of models.

When $\alpha = 0.0$, agents act randomly because they do not learn by either of the learning algorithms. By employing the proposed method, we can identify the distributions of the series data of the stock price; the data are generated by agents who act randomly from the distributions of the stock price series generated by agents who acquire rational investment behaviours.

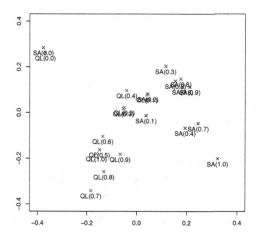

Fig. 3. Result of comparison of Q-learning with SARSA

On the other hand, when $\alpha \neq 0.0$, agents acquire rational behaviours. The simulation results are different because of the use of different learning algorithms and parameter setups. This implies that the influence of the learning algorithms on the simulation results should be investigated before investigating any complex social problems. When α is small, the simulation results are minimally influenced by the learning algorithms used.

4.3 Extraction of Similarity and Dissimilarity Relations by Using Cluster Analysis

In the previous experiment, we identified similar results by employing diagrammatic representations. However, it is necessary to define the criterion for classifying results as being similar. We utilize a cluster analysis for this purpose. Generally, a cluster analysis is employed to classify objects into different groups so that the objects in each group share a common merit. By applying cluster analysis to the distance matrix shown in eq. (5), we can classify multiple simulation results. Thus, to obtain simulation results that are minimally influenced by the learning algorithms, we simply have to extract clusters that include similar results from models using different learning algorithms.

We perform an experiment to evaluate this technique. The experimental setup is identical to that of the previous experiment. The method employed is as follows;

1. Apply hierarchical cluster analysis to the distance matrix, except in the case of $\alpha = 0.0$.
2. Cut the dendrogram at a specific value.
3. Extract clusters that include similar results from models using different learning algorithms.

Figure 4 shows the result of cluster analysis. We cut the dendrogram at a threshold value of Hight = 0.5 and obtained seven clusters. Table 3 shows two

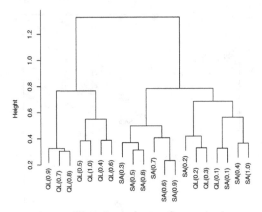

Fig. 4. Result of cluster analysis from models using Q-learning and SARSA

Table 3. Clusters that include similar simulation results from different models

Cluster1	QL(0.1), SA(0.1)
Cluster2	QL(0.2), QL(0.3), SA(0.2)

of those clusters. Thus, we were able to obtain simulation results for small α, similar to the previous experiments. One problem faced in this experiment is the lack of guidelines for selecting a threshold value to cut the dendrogram.

5   Discussion

The proposed comparison method consists of a feature extraction technique and a method to measure the distance between distributions. Since the processes involved in each technique do not depend on the structure of the computational models, our method can be applied to general multi-agent-based models. When classifying simulation results, it is important to use the same criteria that other experimenters generally use for classification. However, there are diverse viewpoints to analyze simulation results, and therefore, it is difficult to find a general one. For example, in the field of quantitative economics, statistical models such as ARMA, ARIMA and ARCH have been developed to analyze time series data [13]. The main purpose of these approaches is not the classification of features but their forecast. Therefore, these techniques are not suitable for our purpose. For the purpose of classification of features, feature extraction techniques have been employed. Since the DFT technique, a popular feature extraction technique, is generally applied to time series data, models can be compared using the same evaluation criteria. As the first step, we generated rough plots of the time series data and compared them. Recently, discrete wavelet transform (DWT) [14] has been gaining popularity. The DWT has several advantages; e.g., although it has a complexity of $O(n)$ and multiresolution property; however, [15] showed that the performances of both transforms do not significantly differ.

In our method, it is not necessary to define the objective function of the model; this is the main difference between our method and Terano's method [16].

6 Conclusion

To expand the application area of model-to-model approaches, this paper introduced a quantitative method for comparing multi-agent-based simulation models that have the following properties: (1) time series data is regarded as a simulation result and (2) simulation results are different each time the model is used due to the influence of randomness, even though the parameter setups in each use are identical. To evaluate the effectiveness of our method, we applied it to an artificial stock market model. From the purpose of cross-element validation, we introduced two different learning algorithms, Q-learning and SARSA. Through an experiment, we showed that the distances between those distributions that contain a specific number of simulation results are stable. Through other experiments, we demonstrated that our method is useful for (1) investigating the difference in the trends of simulation results obtained by using different learning algorithms; and (2) identifying the reliable features in the simulation results that are minimally influenced by the learning algorithms used.

References

1. Takadama, K., Fujita, H.: Toward guidelines for modeling learning agents in multiagent-based simulation: Implications from Q-learning and Sarsa agents. In: Davidsson, P., Logan, B., Takadama, K. (eds.) MABS 2004. LNCS (LNAI), vol. 3415, pp. 159–172. Springer, Heidelberg (2005)
2. Axtell, R., Axelrod, R., Epstein, J., Cohen, M.: Aligning simulation models: A case study and results. Santa Fe Institute, Working Papers 95-07-065 (1995)
3. Takadama, K., Suematsu, Y.L., Sugimoto, N., Nawa, N.E., Shimohara, K.: Cross-element validation in multiagent-based simulation: Switching learning mechanisms in agents. Journal of Artificial Societies and Social Simulation 6(4) (2003), http://jasss.soc.surrey.ac.uk/6/4/6.html
4. Keogh, E., Kasetty, S.: On the need for time series data mining benchmarks: A survey and empirical demonstration. In: ACM SIGKDD International Conference on Knowledge discovery and data mining, pp. 102–111 (2002)
5. Morchen, F.: Time series feature extraction for data mining using DWT and DFT. Technical Report 33. Departement of Mathematics and Computer Science, Philipps-University Marburg (2003)
6. Beyer, K., Goldstein, J., Ramakrishnan, R., Shaft, U.: When is "nearest neighbor" meaningful? In: Beeri, C., Bruneman, P. (eds.) ICDT 1999. LNCS, vol. 1540, pp. 217–235. Springer, Heidelberg (1999)
7. Agrawal, R., Faloutsos, C., Arun, N., Swami, A.N.: Efficient Similarity Search In Sequence Databases. In: Lomet, D. (ed.) 4th International Conference of Foundations of Data Organization and Algorithms (FODO), pp. 69–84. Springer, Heidelberg (1993)
8. Duda, R.O., Hart, P.E., Stork, D.G.: Pattern Classification, 2nd edn. John Wiley & Sons, Inc., Chichester (2000)

9. Torgerson, W.S.: Multidimensional scaling: I. theory and method. Psychometrika 17(4), 401–419 (1952)
10. Eckart, C., Young, G.: Approximation of one matrix by another of lower rank. Psychometrika 1(3), 211–218 (1936)
11. Palmer, R.G., Arthur, W.B., Holland, J.H., LeBaron, B., Tayler, P.: Artificial economic life: a simple model of a stockmarket. Physica D: Nonlinear Phenomena 75(1-3), 264–274 (1994)
12. Arthur, W.B., Holland, J.H., LeBaron, B., Palmer, R.G., Tayler, P.: Asset Pricing Under Endogenous Expectations in an Artificial Stock Market. Santa Fe Institute, Working Paper 96-12-093 (1996)
13. Tsay, R.S.: Analysis of financial time series: financial econometrics. John Wiley & Sons, Inc., Chichester (2002)
14. Chan, K.P., Fu, A.W.C.: Efficient time series matching by wavelets. In: 15th International Conference on Data Engineering, pp. 126–133. IEEE Computer Society, Los Alamitos (1999)
15. Wu, Y.L., Agrawal, D., Abbadi, A.E.: A comparison of DFT and DWT based similarity search in time-series databases. In: Ninth Internatinal Conference on Information and Knowledge Management, pp. 488–495 (2000)
16. Terano, T.: Exploring the vast parameter space of multi-agent based simulation. In: Antunes, L., Takadama, K. (eds.) MABS 2006. LNCS (LNAI), vol. 4442, pp. 1–14. Springer, Heidelberg (2007)

Deepening the Demographic Mechanisms in a Data-Driven Social Simulation of Moral Values Evolution

Samer Hassan[1], Luis Antunes[2], and Millán Arroyo[1]

[1] Universidad Complutense de Madrid, Spain
{samer@fdi,millan@cps}.ucm.es
[2] GUESS/Universidade de Lisboa, Portugal
xarax@di.fc.ul.pt

Abstract. The "Keep It Simple, Stupid" principle is a recommended rule for modelling complex phenomena. However, there must be a compromise between simplification and expressiveness, determined by the results produced by the model. Here we propose to gradually increase the complexity of a model, so we can improve its behaviour. This incremental "deepening" process is an attempt to approach the real phenomena, so resulting in a better model, provided that an accurate analysis reveals the right steps. As application we propose an agent-based data-driven model of the evolution of moral values in the Spanish post-modern society. We focus on improving the demographic mechanisms so that the system output follows the evolution of Spanish population. In order to do that, we raise the amount of quantitative input information of the system, improve its statistical distributions, and change the time of evolution, together with other commented changes.

Keywords: agent-based modelling, agent-based social simulation, complexity, demography, values.

1 Introduction

Simulation for the study of complex social phenomena is often based on multi-agent systems, drawing on micro motives to explain and cause individual decisions and interactions of the participant agents [20]. The role of the experimenter is in this case to design the appropriate simulation setting and to observe the outcomes, both in terms of overall macro behaviours, as well as individual agent trajectories (histories over time) [2]. Many times, this aggregate behaviour is called emergent, as the collective and even individual behaviours could not be predicted or expected (or predictable) from the initial settings of the simulation [4].

The Mentat model [8] takes a relatively different approach from the one described above. In fact, it uses data from a survey conducted over thousands of people across Europe to set up the initial state of the simulation and to justify the behaviour that agents exhibit along the simulation. Instead of using individual motives for individual behaviour, in Mentat the agent heterogeneity

N. David and J.S. Sichmann (Eds.): MABS 2008, LNAI 5269, pp. 167–182, 2009.

is achieved by a thorough examination of the aggregate measures taken over those thousands of agents in the survey. Specific domain knowledge is used to isolate the relevant groups, and then aggregate statistical measures are used to characterise the agents in those groups and their behaviour.

Since the survey has been repeated in time, it was possible to test out the appropriateness of this approach by comparing the survey results of 2000 with the outcomes of 20 years of simulated time starting with real data for 1980. This was made by focusing on the evolution of moral and political values for the Spanish population, and results were quite accurate.

In this paper, we have identified and isolated several issues concerning demographic evolution that seemed to be unaddressed or overly simplistic in the current Mentat version. For each of these issues, we have proposed a solution and tried it out in Mentat in a progressive deepening manner. These mechanisms allowed us to avoid the overall decrease in population numbers, which were not found in the real survey.

The methodology used to improve the model was the incremental deepening of its mechanisms. It will improve the micro-behaviour for a better explanation of the macro-level, in the same line of [5]. However, as a result of its application we don't want to explain the whole process, but the importance of demographic dynamics in the evolution of values trends.

To tackle the initial absence of under-18s in the simulation we have used the real data to generate children, their mental setting, and introduced them early in the simulation. This fills the gap for about 20% of the population that was previously absent.

Marriages were also absent from the initial population, agents would only get married later during the course of the simulation. We have proposed a method for introducing an initial set of marriages, as a more realistic picture of the population.

Another issue concerned reproduction. Where previously only an overall population average was used, we have now considered some mechanisms building on age-based clustering of reproductive women. This allows us to observe the dynamics of reproduction among several groups of women, which is of course a crucial matter in the study of moral and behavioural values.

The paper is organised into 7 sections. Section 2 provides context of the research context for the study of moral values evolution in Spain in recent years. Section 3 details the variables to be studied. Section 4 describes Mentat model as an exploratory social simulation to address the issue at hand. Section 5 begins with the description of our methodology. Afterwards, the following subsections illustrates the demographic mechanisms previously used in Mentat, criticise their adequacy, and propose deeper ones, while we describe the details of the implementation used in the deepened mechanisms, and how they affected the development of the simulation. Section 6 presents and analyses the simulation results, and finally section 7 concludes and proposes some directions for future work.

2 Research Context: The Evolution of Moral Values

Social phenomena are extremely complicated and unpredictable, since they involve complex interaction and mutual interdependence networks. Quantitative sociological explanations deal with large complex models, involving many dynamic factors, not subject to laws, but to trends, which can affect individuals in a probabilistic way. According to [16], a social system is an interrelated and hierarchical set of components which interact to produce certain behaviours. So, we can conceive our target social system as a collection of individuals that interact between them, evolving autonomously and motivated by their own beliefs and personal goals, and the circumstances of their social environment.

The idea beneath Agent-Based Social Simulation (ABSS) is that we may be able to understand this huge complexity not by trying to model it at the global level but instead as emergent properties of local interaction between adaptive autonomous agents who influence one another in response to the influences they receive [11]. Because of that, the specification of characteristics and behaviour of each agent is critical, in what it can affect the dimensions of the studied problem.

As part of the middle-term objective of increasing the usability of ABSS tools for sociologists [18,19], who are usually not skilled in computer programming, we looked for a real sociological issue to analyse. Therefore, the case under study makes an analysis of the evolution of multiple factors in Spain between 1980 and 2000, focusing on moral values and mental attitudes. This Spanish period is very interesting for research, due to the big shift on values and attitudes that the society bore then. The almost 40 years of dictatorship finished on 1975, when the country was far from Europe in all the progress indicators, including the predominant values and modernisation level. However, the observed trends of the values and attitudes evolution since then are analogous to the ones found in its EU partners. Furthermore, the change in Spain has been developed with a special speed and intensity during the studied period. The problem we are facing is to study this complex problem: the shift in the people's mentalities in this society, in this period.

The Spanish society had in 1981 some more traditional and conservative values with respect to the European average, due to the dictator regimen influence. However, in 1999, the measurements exhibit a relatively high permissiveness, showing not only a strong convergence with European values, but also positioning itself as one of the most modern countries [10,15]. The strength and rapid velocity with which these attitude change occurred in the Spanish society imply, as a starting point in 1981, the existence of strong intergenerational differences, that is, between different age groups, which are more intense in Spain than in the other Western Europe countries, as shown by the European Value Survey. The age variable ability to discriminate values and mental attitudes is in general important in all the modern industrial societies such as those from Western Europe, but that ability is especially high in Spain, not only in 1981, but along the whole addressed time period. Hence the importance of demographic dynamics as a predictor of values change during the whole period, especially in this country, which was facing a quick change process.

The problem is faced using ABSS, but which cognitive model should be used for the design of the agents? The widely used Believes-Desires-Intentions architecture (BDI) [3] has been proposed as a possibility, but it has been discarded. The BDI model, as other models based on the rational choice theory, works very well for limited contexts, with clear objectives and roles, together with consistent defined rules. Good examples are the industrial task-driven agents or firms in a perfect market. But we found it useless in this general context, where the agents' global objectives cannot be defined. Besides, the values change process cannot adapt well to the $B + D \rightarrow I$ classical pattern: typically, the change comes through the world circumstances, that change the behaviour (Intentions), and only then, afterwards, the B and D are adapted in consequence. In this case the knowledge and mental state is more a result of the external forces than the opposite.

The values are influenced by a large number of factors. We need to cope with the most important ones: gender, age, education, economy, political ideology, religiosity, family, friend relationships, matchmaking and reproduction patterns, life cycles, tolerance regarding several subjects... factors that are usually inter-related. The statistics of these variables will evolve over time, together with the agent network.

We consider that the best way to deal with this overwhelming complexity is to reduce the number of independent variables to the minimum. But several studies in the field show the difficulty of reducing the number of parameters to only a few, as many other ABSS systems do, studying only particular aspects of the problem. Then, we have decided to solve the problem loading huge amounts of empirical data into the social simulation. This way, the variables loaded are not independent anymore: they are treated as known values. The dynamic processes that are not constant can be controlled and fixed too by empirical data: in this case, *Normal* distributions with known means (average of children per couple, average of female death age...).

Therefore, the initial data for agents of the simulation has been taken from the results of the European Values Survey (EVS). EVS provides a source of quantitative information and periodical results offering also data for validation of the simulation model. EVS is run periodically (each 10 years) in all European countries, regarding a huge number of characteristics. The data is aggregated by country, so we can easily extract the answers of the sample of Spanish individuals. Those 2303 individuals are a good representative sample of the Spanish society, and we can consider that their distributions of the studied variables are equivalent to the real ones.

In the case under study, we will import the data from the EVS of 1981. There is no available data from before that, during the authoritarian regime. The following EVS (1990 and 1999) will be used for validating the results of the system, as "snapshots" of the reality.

3   Variables Studied in the Simulation

The European and World surveys on values confirm an increasing change in the values of developed industrial societies. This has suddenly occurred with strength in the Spanish society during the time period we studied (1981–1999). Inglehart [9,10] has named the macro-trends that define these change in values post-materialism and more recently post-modernisation. Halman [7] has referred to the process of social individualisation. Both concepts are closely interrelated, as well as related to the secularisation process, that has been more intense in Europe than anywhere else in the World [14].

Not in every society we will see the demographic dynamics yielding the good prediction results seen in the Spanish society. In Spain we can see an especially high discrimination ability among the values and attitudes of individuals with several ages, sustained along at least three decades. In other countries this intergenerational differentiation is much smaller, and so is the impact of demographic dynamics. An especially appropriate example is Portugal, that went trough a similar democratisation and openness phase in the same years, but does not have this accentuated differentiation.

Our simulation aims to determine to which extent the demographic dynamics explains the magnitude of mentality change in Spain. The demographic factor does not cause nor determines the change in values, but it does exert an important influence on the velocity and intensity in which it manifests, so (and for this reason) it possesses an important predictive ability for its evolution, as we will show further on. This is mostly due to the fact that the changes in values have been chiefly (but not exclusively) generational changes, hence the generational replacements (say, the death of elders, carriers of the most traditional and conservative values, and the arrival of youngsters, bearers of emerging values) constitute a significant sociological inertia.

Therefore, the values of each individual remain constant, but not their aggregation in the whole society, as its demography is changing with time. This reflects the inter-generational changes, but not the intra-generational ones, also important enough. However, this isolation is the only way to analyze and appreciate the predictor effect of the demographic dynamics.

The variables chosen for mentality change indicators, attached to the macro-tendencies of individualisation, postmodernisation and secularisation have been, on one hand, a set of signals for individual moral tolerances: abortion, divorce, euthanasia and suicide, and on another, a signal for religiosity, which classifies the several degrees of religiosity of the population in the categories of: clericals, low-intensity religious, alternative-critical with church hierarchy, and finally 'not religious.'

Furthermore, we exerted control to ensure that the evolution of demographic variables provide an adequate response (we provide data relative to the most important: gender and age). We also considered other variables, such as the socioeconomic status, an indicator of education level, and another of ideological positioning. The latter variables are not the focus of study, nor they are controlled in the model, but were nevertheless considered to verify the degree to which the simulation was realistic in this other social and socioeconomic dimension. The indication of education level is defined by the age in which school was abandoned as a main activity. The indicator of social status is a standardised factor (average zero and standard deviation one), and ideally it should stay (close to) constant in time. The indicator for political ideology was obtained from a scale of ideological self-positioning in the spectrum of leftwing-rightwing positions: 1 means extreme leftwing and 10 means extreme rightwing.

4   MENTAT: Data Driven Social Simulation Model

In previous works, we can see a progressive increasing permissiveness and moral relativism, an important religious secularisation and a certain decrease in ideological dogmas, concomitantly with an increase in the 'postmodern' [10] and individualising [7] sensibilities. Reciprocally, these trends are interconnected among them. On the other hand, for the purpose of this study, it is especially relevant to show that, according to Ronald Inglehart [9,10], basic values are acquired in a relatively early stage in life, between the adolescence and the first youth, and these will change very little later in life. This is the theoretical principle that this prominent sociologist uses to explain the fact that the values changes that occurred in Spain in the second half of XX century and until now, are indeed intergenerational changes. The explanation of the changes arises from the environmental conditions in which the socialisation of the new generations occurs.

Having in mind the generational differences so accused in Spain, in what respects to values and attitudes, from the previous theory we infer that the demographic dynamics in Spain should reach a high predictive value of the prediction of such attitudes for several decades, since the generational gaps will tend to keep along time. A way of showing that this theory is right is to study the effect of demographic dynamics without taking into account the intergenerational changes. If the theory is right, we should observe that the simulation data from 1981 should adjust relatively well (even when time is taken into account) to those obtained empirically in the later curves of 1991 and 1999.

As it has been mentioned, the methodology of research chosen will be the ABSS, taking care in detail of the underlying sociological model, based on the works of studies of moral values of [12,13]. A first prototype of MENTAT was implemented, as explained in [17] and [8]. What is presented here will try to improve the results of these other works.

The Multi-Agent System (MAS) designed was developed using agents with several attributes: from the most simple ones such as sex or age, to others like economic class, ideology, or divorce acceptance. The attributes have been carefully selected as the most important for our purposes: the evolution of moral values. The population in the agents society (as in real societies) also experiments demographic changes: individuals are subject to life cycle patterns: they are born, can find a couple, reproduce and die, going through several stages where they follow some intentional and behavioural patterns. Besides, the agents can build and be part of relational groups with other agents. They communicate with their wide Moore neighbourhood, and depending on their rate of similarity, occasionally leading to friendship relationships. On the other hand, after finding a couple they can build family nuclei as children are born close to their parents.

The Mentat system may be configured to follow the parameters (such as average number of children per couple, or mean of male average age of death) from a specific country or import data from surveys that specify the attributes of the agents, reflecting the behaviour of the given population (and as mentioned, the EVS will be imported).

Besides, due to the relative simplicity of the agents, it can manage thousands of them, reaching the necessary amount for observing an emergent behaviour that results from the interactions of individuals, leading to the appearance of social patterns than can be studied [2]. Study is helped by the capacity of plotting several graphs, during and after the execution of the simulation, which reflect the evolution of the main attributes of the social system (some will be shown later on). These possibilities are possible thanks to the potential of Repast, the leading framework in java programming for social simulation. The system robustness has been tested enough to demonstrate the stability of the results, needed for the macroscopic comparative analysis.

Now we are focused in the study of the influence of demographic dynamics in inter-generational change. Therefore, not intra-generational evolution has been implemented yet: the agents' mental states don't evolve internally in time (they will always have the same level of education, ideology, or acceptance of divorce). The global evolution of the variables that can be observed is due to the changes on the demography (the old agents are dying, new ones are born after reproduction, but they have very different values).

Thus, in the micro level the agents' interactions were modelled, representing the socialisation processes. Note that these processes will not be completely represented until the model includes the intra-generational dynamics, with the influence between agents and so, values diffusion. Now they are limited to establishing friendship and couple links with the aim of reproduction, together with the family links. In the meantime, the evolution of the different patterns can be observed on the macro level.

Taking the EVS of 1981 of Spain, a sociological approach provided a spreadsheet for the characterisation of the group of individuals which statistically represents the Spanish population. In the original system, only 500 individuals were managed. However, nowadays Mentat is able to deal with all the individuals presented in the Spanish part of the EVS: 2303 individuals loaded as the same number of agents. This relation one-to-one allows a better behaviour.

Thus, these data were taken as input to generate the diverse and representative population of agents in the model, which was simulated for a period of nearly 20 years, till year 1999. As it has been mentioned, the evolution of the big number of variables studied can be compared with the data of the other two EVS, validating the results of the ABSS.

Note that as the system is non-deterministic, the graphical results have some variations at each execution. The outcome of the model should not be taken as a static output. In all our simulations the trends were always very similar, even though the exact data have some small comparison errors. Therefore, the system executions have structural similarity, as defined in [6].

It has to be remarked that, as long as we are analysing the evolution of moral values and socio-cultural phenomena (deeply interrelated) during a long period of time (like 20 years), and we pretend to achieve a simulated output in equivalent quantitative terms to their real evolution, it becomes a need to implement an analog demographic pattern of the Spanish one. And we will proceed with this task in the following section.

5   Demographic Mechanisms

5.1   Incremental Deepening as a Methodological Approach

When building up experimental designs, it is usual to defend and adopt the so-called KISS ("keep it simple, stupid!") principle [2]. In some sense, Sloman's "broad but shallow" design principle starts off from this principle [21,22]. Still, models must never be simpler than they should. The proposed solution for this tension is to take the shallow design and increasingly deepen it while gaining insight and understanding about the problem at hand. The idea is to explore the design of agents, interactions, institutions, societies and finally experiments (including simulations and analysis of their outcomes) by making the initially simple (and simplistic) particular notion used increasingly more complex, dynamic, and rooted in consubstantiated facts (see [1]). As Moss argued in his WCSS'06 plenary presentation, "Arbitrary assumptions must be relaxed in a way that reflects some evidence." This complex movement involves the experimenter him/herself, and according to Moss includes "qualitative micro validation and verification (V&V), numerical macro V&V, top-down verification, bottom-up validation," all of this whereas facing that "equation models are not possible, due to finite precision of computers." Therefore, this paper will follow the same line of his paper [5].

A possible sequence of deepening a concept, representing some agent feature, (say parameter c, standing for honesty, income, or whatever) could be to consider it initially a constant, then a variable, then assign it some random distribution, then some empirically validated random distribution, then include a dedicated mechanism for calculating c, then an adaptive mechanism for calculating c, then to substitute c altogether for a mechanism, and so on and so forth.

The deepening procedure is meant to be used in a broad methodology in which the space of possible designs for agents, societies and experiments is explored through a combination of techniques that allow to discover the most adequate features of models, having in mind the increase of insight that the models and their explorers possess. Some of the ultimate goals of social simulation include the explanation of phenomena, prediction of future outcomes, and even prescriptions to tune up the effects of policies. By traversing the design space, the best designs can be mixed and used to develop more accurate models to fulfill those purposes.

5.2   Introducing Missing Children

The initial demographic pattern that we have is a probabilistic one, with a representative distribution given by the EVS of 1981 (where the maximum random error limit is $\approx 2.08\%$, the trust level is 95.5% and $p = q = 50\%$). This distribution represents the Spanish population, with the adequate distribution in many different variables: education, economy, ideology, values, etc. But there is a methodological problem difficult to face: as it is based on a survey, we found no data from underaged people. Children don't make surveys, and the problem is not the limit of age, because we will always find a small range of age that is not able of giving us the appropriate answers. The problem could be ignored if we consider just the very short-term. But dealing with 20 years evolution, the problem is deep and important: in 1999, all the children in 1981 are adults (even the ones of 0 years, that in the end are 18) capable of reproducing and altering the global patterns. Furthermore, those "children" (some of them have 17 years, that would very likely reproduce in the 80's) represented the 23.33% of the total Spanish population at that time... a too big amount for being ignored, especially considering that lots of them will reproduce during those 20 years. As a side effect of these "missing children" (and as important as that: the missing children that should be born from them but are not) the population of the simulation drops more than a 20% during those 20 years (a whole old generation dies, but only a few children are born).

So the first task has been to introduce that big amount of individuals that we do not have data from. The characteristics of the underaged born after 1981 are based in the EVS-1981. Although the initial idea was to use the information of the EVS-1990 to generate them, it was discarded because of several reasons. A big part of them (the closest to 18 yeas) were much more similar to the 1981 group than to the 1990 generation. Besides, the 1990 data would give individuals already changed by the influence of other circumstances, and we want to study the influence of demography in these variables. This way we will still use the other two EVS just for validation, and the work will continue being general enough (considering that, in many cases when trying to make predictions, we don't have available "future data" compared with the initial time).

As we have available the population pyramids for these periods (in the official statistics), we can easily know how many individuals of each range of ages we need: in total, 716 new agents that are included. Their characteristics have been assigned from other existing agents of the EVS-1981, chosen randomly from the ones under 30 years. This way of proceeding is justified in that the children should be similar to the youngest ones available. We are aware that in the real society this new generation keeps values slightly more modern than the simulated ones, but with the options available, it turns to be an acceptable approximation for solving the problem without an exponential growth of complexity.

But solving this demographic problem, including the underaged with a group of characteristics equivalent to an adult, raises two new inconveniences. First, it disrupts reality as long as children are not mature enough for having stable values. Second, it makes it difficult to compare with the available empirical data (basically, surveys and studies made after 1981), always done only to people over 18 years. Both problems can be solved filtering the output of the simulation: the ones related to the variables measured (as if we were doing a survey in the actual population of the simulation). Anyway, some demographic statistics will remain unfiltered (total number of agents, percentage of children and adults, and others).

5.3   Introducing Initial Marriages

After the "jump" from the Mentat with (around) 2300 individuals to the Mentat 3000 (being precise, we should have called it Mentat 3019), we still found structural deficiencies concerning the demography management in the system. It can be observed that in the first years of simulation no agents are born. Obviously, this is completely unrealistic, and the reason of this misbehaviour is that the agents begin isolated from each other (close, but with no links between them). They invest their first years in the simulated world in finding friends, and maybe a couple. When the population as a whole has already established a robust linked network, they begin showing the expected macro output.

Then, for dealing with this small difficulty we have to introduce the needed initial marriages, as we did with the needed initial children. But, even though we have the percentage of people that should be married (and even the agents that are married or not, as it is given in the EVS), we can't know with whom (it depends on the network distribution, as an agent only have communication with his local environment).

As we couldn't load this missing information, and although we could have forced the links by inventing them, we decided to let the agents decide with their criteria. First, we loaded the information available: which agents are married/single in 1981. Second, we let the simulation begin... but freezing the years counter. In this special period (named "Phase B"), the agents neither get older, nor have children, nor die. But they do communicate with each others, building new friendship and couple links (taken into account if they should be married in 1981). After a certain period of "steps" (the minimum measure of simulated time) have passed, the "Phase B" finishes and the years counter begins. This way, the real simulation begins with the agents already linked, and since 1981 new agents are born, achieving a much more realistic global behaviour.

5.4   Deepening Population Dynamics

With the last changes, new dilemmas arise. In the previous Mentat 2300, before all these new changes, the agents only searched for a couple whenever they wanted to reproduce, so the number of total couples was the same as the number of couples with children. But now, taking into account the agents already married before 1981, this doesn't make sense. The most part of those old couples probably had already all the children they wanted to have, and shouldn't have more in the 80's and 90's. So we need a way to find out who wants to have more children.

This fact is determined by multiple factors (economic class, religiosity...) but the one that beats them all as the biggest weight in the election/possibility of having children is the age. With the data available, and with the statistics technique of nonlinear regression, we can build an equation that will allow us to determine the probability of having children of a given individual of certain age. For example, an agent of 35 years old has only a 23% of probability.

The same mechanism can be used for other needed choices. Instead of loading the information of being single from the EVS (as mentioned in the last subsection), we can calculate it through another regression equation, so it is generalised and usable for every agent (and not only for the loaded ones).

To continue to try to achieve the demographic convergence with the target, we decide to substitute the quite simple way of deciding the number of children (a Normal distribution with centre on the Spanish average in 1981) with the regression equation of birth rate considering the actual year. The change on the behaviour is immediate, as the birth rate has decreased rapidly in those years (from a 2.2 falling to 1.19). It has to be compensated with the introduction of a proper distribution for life expectancy (by the way, two equations, as they are completely different for men and women), replacing the previous Normal one. In this case, it has been increasing thanks to the improvements in health and quality of living. Lastly, we cannot ignore that the age of mothers when having the first child has been increasing significantly. Therefore, we exchanged the last normal distribution for another appropriate regression equation based on the data available.

Now the life cycle of an agent has changed, being more complex, but still easily understandable. Its much richer in information, as it uses more data from the EVS, together with some equations (generated from empirical data). However, this solution has its own limitations. The first two equations, based on age, try to solve the problem of "time," as an agent has different behaviour depending on his/her age. But the equations themselves evolve in time too. Maybe that agent of 35 years old has a 23% of probability in 1981, but a 28% in 1995. Anyway, the calculations would turn to be thorny and unnecessary complicated for the small real difference that it would entail. And the other equations, that take into account the year instead of the age, do not imply this difficulty.

6 Results and Discussion

In table 1 we can observe a comparison between the different evolutions of a sample of chosen variables, in the period studied. The measurements have been a) made by extracting the information from the three different EVS of those years, b) from statistical calculations (over several executions) in the previous version of the system, Mentat 2300, in the specified years, and c) equivalent calculations made in the new Mentat 3000. All variables are calculated considering only the individuals over 18 years. We extracted a wide range of different statistical measures to test the consequences of every change. Even so, we are showing here mainly means and percentages, as they can easily reflect the behaviour of the whole population. It is important to mention that the high stability of both versions of Mentat has simplified the analysis, and we can conformably assess that these values have a minimum error between executions.

We can begin the analysis of the results by the most simple measure: the gender. Its evolution is steady, but it has a small change in 1990. This change

Table 1. Validation: comparison between EVS, previous version and the more complex one

	EVS			Old Mentat 2300			New Mentat 3000**		
	1981	1990	1999	1981	1990	1999	1981	1990	1999
GENDER									
Men	49	47	49	49	48	48	49	48	49
Women	51	53	51	51	52	52	51	52	51
Age (mean)	45	43	46	45	51	57	45	47	49
% 65+ years	15	13	19	15	23	31	15	19	24
% 65+ years*	16*	18*	21*						
% SINGLE	28	29	29	100	82	79	100**	42**	35**
% SINGLE***							100***	34***	30***
% SINGLE****							100****	29****	28****
AGE END OF STUDIES	16	N/A	17	16	16	16	16	18	18
ECONOMIC STATUS	0	N/A	N/A	0.028	0.062	0.095	0	0.09	0.105
IDEOLOGY (mean)	4.85	4.65	4.75	4.85	4.82	4.73	4.85	4.74	4.59
IDEOLOGY (%)									
Left	29	33	31	29	30	32	29	33	36
Centre	18	19	23	18	17	18	18	18	17
N/A	30	25	24	30	30	29	30	29	27
Right	22	23	21	23	22	22	23	22	20
TOLERANCE MEASURES (scale 1 to 10)									
Abort	2.89	4.43	4.58	2.89	2.96	3.06	2.89	3.08	3.3
Divorce	4.79	5.65	6.24	4.79	4.92	5.09	4.79	5.13	5.4
Euthanasia	3.18	4.17	4.95	3.18	3.24	3.34	3.18	3.43	3.6
Suicide	2.26	2.25	2.95	2.26	2.29	2.35	2.26	2.36	2.5
RELIGIOUS TYPOLOGY	1981	1990	1999	1981	1990	1999	1981	1990	1999
Ecclesiastical	33	25	22	33	31	29	33	29	25
Low-Intensity	22	26	23	22	23	22	22	23	22
Alternatives	14	17	19	14	14	15	14	16	16
Non-religious	31	32	35	31	31	33	31	34	37
POPULATION GROWTH			+8%*			+1%			+7.2%**
POPULATION GROWTH***									+8.6%***
POPULATION GROWTH****									+10%****

*: The source is the Spanish population census of the years 1981, 1991 and 2001 (INE, Spain). EVS does not shows accurately these data.

,*,****: With a "Phase B" of 100 steps, 500 steps or 1000 steps, respectively

is not shown at all in the old system, but can be appreciated quite well in the new one. On the other hand, the age mean increases just a bit (+2%), but in both Mentat's we see an incorrect higher rising. Anyway, it's clear that the old version has a much worse output (+22%) than the new one (+8%).

To analyse the percentage of old people has more importance than the previous variables. Its evolution reflects an important part of the population pyramid structure, it is more sensible than the others, and it suffers bigger changes. The EVS data is not accurate here, because it takes into account other factors, but in the table it has been included the data from the census as empirical source. The increase, as before, is ridiculous in the old case, but moderately good in the new one.

The percentage of single individuals in the population is a factor related to the agents network. A single individual is the one that or a) does not have single adult opposite-sex friends to have as a couple, or b) it does not want to have a couple (for example, because it is a child). In the beginning, the agents find the problem (a), but with time the network should grow in complexity and cohesion, so the predominant problem is (b). We can see that the percentage of singles should remain quite stable near 30. However, in both Mentat's we see a surprising "100" in the beginning. This can seem weird, as in all the other measurements the 1981 value matches the EVS. The logic beneath this is that the measurement of 1981 has been done in the very beginning of the simulation, even before than "Phase B" (the introducing of friendships and marriages before the years counting). And so, in that moment, all the agents are isolated, so all find themselves with the problem (a). Obviously, the size of the "Phase B" is crucial for this variable: the more time you leave them to interact, the more couples you will have. We will leave apart the Mentat 2300, with useless output, and concentrate in the new one. We have tested with three different sizes: 100, 500 and 1,000 steps (one year, when they are been counted, is 50 steps), and the different outputs are in the table. After the "Phase B" the network acquires consistence and approaches a lot to the ideal value (that is just measuring the connectivity between nodes). We can see how a wider size gives more cohesion: with 500, we achieve the ideal after 1500 steps (500 of "Phase B" plus around a thousand of the nearly 20 years of simulation), and logically it can be achieved before with 1000, after the same 1500 steps (1000 of "Phase B" plus 500 of 10 years). The stability after that objective is achieved (in 1999 it is still nearly 30, instead of continuing the falling) matches accurately the observed reality, and can lead us to the situation where the problem (b) is the widely dominant. From the Social Network Analysis point of view, we could say that this is a property of small-world networks.

On the other hand, the size of "Phase B" has another logical effect: the more couples we have, the more children will be born. Anyway, as the most part of the young couples find someone always, it does not have a big influence in the population growth. As we can check in the end of the table, the real growth in that period is around +8%. The previous model fails again, but now, with different sizes, we achieve much better results. The side effect of the extra 1000 steps that gave great results with the percentage of singles yields a bigger error here. It has to be mentioned that there are no other important side effects, so we have not shown the other variables with other sizes different than the usual one of 100 steps.

The level of studies shows a slight rising that is approximately shown in the simulation. But the education is categorised (in five ranges), so it cannot be appreciated properly (only with the "jump" from 16 to 18). The economic status is not calculated properly in the EVS, but it is known that should remain stable or with a minimum increase, and this is exactly what happens in the simulation.

The political ideology follows a similar trend but with some more slope, in the means and in the different percentages. The tolerance measures are in the same case: slower increase than in reality, and in both measures the differences between the two ABSS are not particularly big. This is due to several facts. First, we have not modelled the intra-generational changes, so the agents main attributes remain static over time (and these variables are extremely sensible to those influences). Second, the new generation introduced (the 700 agents) have a very similar characteristics to the ones from 18 to 30 (as it has been explained), but they should be more modern than that. Finally, the simulation is not able to display the slight move to the right that occurred in the Spanish society during the governments of Joseé María Aznar (1996-2004). But that would be too much to ask from a simulation drawn from 1981 data, while Spain was still in the period of democratic transition.

One of the best indicators for the evolution of values that we have available here is the religious typology, strongly based on them. As we can see, the values are predicted with a very good accuracy, regardless the different curves that each type follows (rapid fall, hill, smooth rising and smooth growing, respectively).

7   Concluding Remarks

Overall, we can say that we accomplished our objective of improving the results of Mentat through the deepening process. The Mentat 2300 gives quite good trends in some important aspects of the simulation, like the evolution of the religious typology or the political ideology. But, as it has been commented, it has deep problems dealing with the demography of the agents population. With Mentat 3000 we have addressed these problems, keeping the good results that Mentat 2300 achieved, and improving a collection of other indicators. This methodology has been followed, isolating every part that should increase its complexity, re-implementing it, analysing the result of the step, and comparing it to the previous situation and the reality. The result is a model that is still quite simple, especially considering the problem studied. But it has been growing in complexity gradually from the previous version, so now it can deal better with several issues. We have successfully tried to combine simplicity and expressiveness.

As a sociological conclusion, we can extract that the determinism of demographic dynamics in the prediction of social trends is, as we can support by the good results, far more important than what it was expected. Further research should be addressed in this direction, better with other completely different environments that could support the same statement.

After the analysis of the current variables, we suggest that the use of Social Network Analysis (SNA) techniques in combination with the deepening process could be very helpful, as the monitoring of the consequences of every design option could be tracked. Ideally, with automatic testing programs that did not require lots of computational effort, each execution could be deeply analysed and compared with all the previous ones, to build a tree of changes with its SNA measurements associated.

Acknowledgments. This work has been performed in the context of the project "Metodos y herramientas para modelado de sistemas multiagente", supported by Spanish Council for Science and Technology, with grant TIN2005-08501-C03-01.

References

1. Antunes, L., Coelho, H., Balsa, J., Respício, A.: e*plore v.0: Principia for strategic exploration of social simulation experiments design space. In: Takahashi, S., Sallach, D., Rouchier, J. (eds.) Advancing Social Simulation: the First World Congress, pp. 295–306. Springer, Kyoto (2006)
2. Axelrod, R.: Advancing the art of simulation in the social sciences. In: Conte, R., Hegselmann, R., Terna, P. (eds.) Simulating Social Phenomena. LNEMS, vol. 456. Springer, Heidelberg (1997)
3. Bratman, M.E.: Intentions, Plans and Practical Reasoning. Harvard University Press, Cambridge (1987)
4. Conte, R., Castelfranchi, C.: Cognitive and Social Action. UCL Press, London (1995)
5. Edmonds, B., Moss, S.: From kiss to kids an 'anti-simplistic' modelling approach. In: Davidsson, P., Logan, B. (eds.) MABS 2004. LNCS, vol. 3415, pp. 130–144. Springer, Heidelberg (2005)
6. Gilbert, N., Troitzsch, K.G.: Simulation for the Social Scientist, 1st edn. Open University Press (April 1999)
7. Halman, L., Ester, P., de Moor, R.: The individualizing society. Tilburg University Press, Tilburg (1994)
8. Hassan, S., Pavón, J., Arroyo, M., Leon, C.: Agent based simulation framework for quantitative and qualitative social research: Statistics and natural language generation. In: Amblard, F. (ed.) Proceedings of the ESSA 2007: Fourth Conference of the European Social Simulation Association, Toulouse, France, pp. 697–707 (2007)
9. Inglehart, R.: Culture shift in advanced industrial societies. Princeton University Press, Princeton (1991)
10. Inglehart, R.: Modernization and postmodernization. Cultural, economic and political change in 43 societies. Princeton University Press, Princeton (2001)
11. Macy, M.W., Willer, R.: From factors to actors: Computational sociology and agent-based modeling. Annual Review of Sociology 28, 143–166 (2002)
12. Arroyo Menéndez, M.: Cambio cultural y cambio religioso. Tendencias y formas de religiosidad en la España de fin de siglo. Ed Complutense. Servicio de Publicaciones, Madrid (2004)
13. Arroyo Menéndez, M.: Hacia una espiritualidad sin iglesia. In: Tezanos, J.F. (ed.) Tendencias en identidades, valores y creencias, pp. 409–436. Fundación Sistema (2004)
14. Arroyo Menéndez, M.: La fuerza de la religión y la secularización en Europa. Iglesia Viva 223 (2005)

15. Arroyo Menéndez, M.: Individualización y religión en la Europa católica. Revista Española de Sociología (RES) 9 (2008)
16. Parsons, T.: The social System. Free Press of Glencoe IL, New York (1951)
17. Pavón, J., Arroyo, M., Hassan, S., Sansores, C.: Agent based modelling and simulation for the analysis of social patterns. Pattern Recognition Letters (June 2007)
18. Sansores, C., Pavon, J.: Agent-based simulation replication: A model driven architecture approach. In: Gelbukh, A., de Albornoz, Á., Terashima-Marín, H. (eds.) MICAI 2005. LNCS, vol. 3789, pp. 244–253. Springer, Heidelberg (2005)
19. Sansores, C., Pavon, J.: Visual modeling for complex agent-based simulation systems. In: Sichman, J.S., Antunes, L. (eds.) MABS 2005. LNCS, vol. 3891, pp. 174–189. Springer, Heidelberg (2006)
20. Schelling, T.C.: Micromotives and Macrobehavior. W. W. Norton & Company (October 1978)
21. Sloman, A.: Prospects for ai as the general science of intelligence. In: Proc. of AISB 1993. IOS Press, Amsterdam (1993)
22. Sloman, A.: Explorations in design space. In: Proc. of the 11th European Conference on Artificial Intelligence (1994)

Cross-Disciplinary Views on Modelling Complex Systems

Emma Norling[1], Craig R. Powell[2], and Bruce Edmonds[1]

[1] Centre for Policy Modelling
Manchester Metropolitan University
norling@acm.org, bruce@edmonds.name
[2] Theoretical Physics
The University of Manchester
craig.powell@manchester.ac.uk

Abstract. This paper summarises work within an interdisciplinary collaboration which has explored different approaches to modelling complex systems in order to identify and develop common tools and techniques. We present an overview of the models that have been explored and the techniques that have been used by two of the partners within the project. On the one hand, there is a partner with a background in agent-based social simulation, and on the other, one with a background in equation-based modelling in theoretical physics. Together we have examined a number of problems involving complexity, modelling them using different approaches and gaining an understanding of how these alternative approaches may guide our own work. Our main finding has been that the two approaches are complimentary, and are suitable for exploring different aspects of the same problems.

1 Introduction

The NANIA (Novel Approaches to Networks of Interacting Autonomes) project has brought together researchers from a variety of fields who are studying complex systems in which stability, robustness and fitness for purpose emerges through the interaction of evolving, interacting networks of diverse 'autonomes' (where 'autonome' is a term adopted by the project to refer to any interacting multi-state system which can encompass cellular automata, agent, organisms or species). The aim has been to determine global principles with which to describe or control complex autonome systems – an ambitious goal, which was acknowledged from the start was only likely to be partially fulfilled.

In this paper, the work of two groups within the project is examined, particularly the attempts that have been made by the group members to tackle common problems. On the one side, we have a group with a background in agent-based social simulation. On the other, a group with a background in theoretical physics. The application areas studied by both groups are diverse, and the approaches to modelling have been quite different, with one group starting from the agent-based approach, and the other from an equation-based, or system dynamics, approach. Each of these approaches has its benefits and limitations, as will be

N. David and J.S. Sichmann (Eds.): MABS 2008, LNAI 5269, pp. 183–194, 2009.

discussed in this paper. In addition to these two approaches, we discuss a third approach that has been taken, which here we call "individual-based modelling." "Individual-based modelling" as we have termed it does not involve keeping track of individual entities in the way that agent-based modelling does, but is a *discrete* equation-based approach (as opposed to the physicists' original equation-based approach which uses continuous values). This approach has been adopted in an attempt to bridge the gap between the approaches of the two groups.

The two groups entered the project each with a particular application area of interest: in the case of the social simulators, tag-based cooperative groups, and in the case of the physicists, food web evolution (which are explained in Sections 2.1 and 2.2 respectively). Both of these applications are examples of complex systems, in that the behaviour of the system as a whole is strongly dependent on the interactions between individual entities within the system. It is possible to predict certain general properties of the system, such as the *types* of patterns one would expect to see at the system level, but predicting the *actual* system state at a given time is not possible. The purpose of models of these systems is thus not in predicting future states *per se*, but to gain an understanding of how different variables affect certain aspects of system behaviour. For example in tag-based systems, the average level of cooperation is a measure of interest, and in food webs the sensitivity to invaders (new species) is one.

Interactions between the two groups during the project has led to different modelling techniques being applied to the same problems, as part of the process of knowledge sharing. This paper focuses on these shared exercises, discussing the different modelling techniques, the shared problems, the benefits and pitfalls of the different techniques, and the lessons learned. We start in Section 2 by presenting three key problems that we have looked at, then discuss each of the three modelling approaches (equation-based, agent-based and individual-based) in Section 3. We then present a discussion of our shared experiences in Section 4, followed by our conclusions.

2   A Medley of Models

A number of models have been developed by these two partners, some in isolation, and others in collaboration. In this section three key problems are presented: the first two being the starting points for each of the partners; the third was chosen as a problem that was of interest to both groups. Some details of implementation may be found in the following sections, but as we are limited by space we urge interested readers to follow references or contact the authors directly for full details.

2.1   Tag-Based Cooperative Groups

Tag-based cooperative group formation was the original focus of the social simulation research group in this project. Tags are properties of an agent that are observable to others but not hard-wired to its behaviour. Initially suggested by

John Holland [1] they allow (under suitable conditions) the self-organisation of selfish agents to maintain relatively high levels of cooperation.

Tag-based cooperation has been explored in two scenarios as part of this project. The first is in a world of 'symbiotic' sharing, in which agents have the ability to harvest only a subset of the resources that they require in order to exist, and rely upon the generosity of their peers to donate additional resources [2]. Further work has focused on the SLAC (Selfish Link-based Adaptation for Cooperation) algorithm for tag-based cooperation [3], which is used for tasks in which agents must rely on their neighbours for assistance in completion. Rather than using tags in the traditional sense, this algorithm works by allowing agents to observe the behaviour of other agents and some measure of their success, and then use this information to adapt their behaviour following a particular set of rules.

Both of these systems have been shown to achieve periods of very high levels of cooperation, although they also exhibit critical states in which almost nothing is achieved. Variants on the SLAC algorithm have been explored looking to improve system performance [4,5] (with limited success), and ongoing work seeks to explore tasks which involve chains of cooperative agents, rather than simple pair-wise cooperation.

2.2   Evolution of Food Webs

Food web evolution is one aspect of this project that has been discussed before in the MABS forum when in 2006 Norling contrasted two different approaches to modelling the problem [6].

Traditional models of predator-prey relationships, such as the Lotka-Volterra equations, focus narrowly on single predator, single prey relations. Natural ecosystems however typically involve a large number of species: one hundred and eighty-two were identified in what is widely regarded as the most comprehensive study to date: that of Little Rock Lake, Wisconsin [7]. In these large food webs, a species may be basal, in which case it 'feeds' solely from the environment, but may have multiple species preying upon it; it may be a top species, having no predators but possibly several species upon which it preys; or it may be intermediate, in which case it would have one or more each of predator *and* prey species. Figure 1 illustrates an example simplified food web of this type. These different possibilities for interaction give rise to a dynamic *network* as well as population, where the dynamics of the population affects the structure of the network and vice versa.

Although several models of network structure of these large food webs have been proposed (as summarised by Dunne [8, Box 1]), these have focused on *static* structures, not incorporating the dynamism of natural webs. In particular, mutation in natural environments leads to new species in the web, which can have dramatic effects on the overall web structure. The physicists involved in this project have developed a series of models that attempt to capture the evolving structure of these webs over time [9,10,11,12], and members of the social simulation group have attempted to produce an equivalent agent-based model [6].

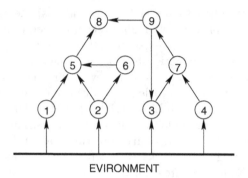

Fig. 1. An illustrative example of a food web. Species 1, 2 and 4 are basal species, 3, 5, 6, 7 and 9 are intermediate species, and 8 is the sole top species.

Extensions to Food Web Evolution. A significant simplification in the food web model described above is its assumption of spatial homogeneity. If territoriality and the boundaries of species range (for example due to altitude) have observable effects on food web structure, these will not be discernible in the existing models. Furthermore, while the models above do incorporate mutation, *invasion* of the web by a new species is a slightly different proposition, as in the latter case the new species may be significantly different from existing species, rather than a slight variant. Extensions to the above work have looked at these issues.

2.3   Opinion Dynamics

Opinion dynamics refers to a wide range of problems, where there are a set of individuals each of which has a particular opinion (from a given range) at any one time. The interactions of members of the group lead to changes in the opinions over time.

The original and best known models in this area are the family of models starting with Deffaunt et al [13]. In these models, agents interact in a pairwise fashion (either randomly chosen pairs or restricted by a given social network) with the opinion of one node affecting that of the other to make it more similar to its own, and having a greater impact the closer or more coherent are the two individuals' opinions. The overriding dynamic in such models is the clustering of individuals into 'groups' with similar opinions – the key questions being how many clusters form, how long does it take and how stable they are.

Departing from the spirit of Deffaunt et al's models, the teams developed a new simple starting point model. This had a fixed number of nodes and directed arcs between these. Each node has a numeric value representing the strength of its belief on a certain issue as well as a value representing its susceptibility to influence by another. During each iteration of the simulation, a random arc is selected and the the opinion of the node at the origin of this is copied into that of the destination with a probability of its susceptibility. Thus eventually (without noise and given the network is connected) all nodes will have the same opinion and change will cease.

A second related model was an already investigated model of phoneme change. This had the structure of the one described immediately above, where the opinions were of the form of a discrete probability distribution representing the chance of uttering each of a set of pronunciations of a word. In each pairing the origin node makes a number of utterances randomly selected according to the distribution. Each utterance is heard by origin and destination node, and changes or updates those probabilities such that those heard more often become higher. This model has been used to try and estimate parameters for a notable case study - the emergence of a coherent New Zealand accent [14]. Very general analytic results about the dynamics of this model are reported in [15].

A third model was specified with a view to modelling the process of consensus formation among a population of separately reasoning agents connected in a social network [16]. Here the opinion of each agent can be thought of as a binary string, where each bit indicates the belief (or not) of each of a sequence of possible beliefs. There is a consistency function from possible bit strings to the interval [-1, 1] that indicates the consistency of this set of beliefs. The copy process involves the copying of a single bit from origin to destination according to a probability determined by the change in consistency that would result in the destination node. There is also a 'drop' process done by single nodes which may drop a belief according to a probability related to the change in consistency that would result from this. Thus it is much more likely that the consistency of the beliefs in each node will increase over time, and also that nodes with similar beliefs will be clustered together.

3   Three Different Approaches to Modelling

As previously mentioned, between the two groups we have tackled the above problems (and others besides) using three different types of modelling. Here we explain how these approaches work, using the food web evolution problem as an example. In the following section, we discuss the issues that have been raised by the different approaches and the ways in which the two groups have influenced each other.

3.1   Equation-Based Modelling

Equation-based simulation, or system dynamics models, attempt to capture the behaviour of a system with a set of equations. These equations are derived from an understanding of individual behaviours, but there is no notion of individuals within the models, and variables are typically continuous valued.

Take for example the equation governing short-term population dynamics in the food web model of [9]:

$$N(i, t+1) = \gamma_{i,0} R + \sum_j \gamma_{i,j} \lambda N(j, t) - \lambda N(i, t) \sum_j \gamma_{j,i} \tag{1}$$

where $N(i, t)$ is the number of resources (assumed to be equal to the number of individuals) of species i, R is the number of environmental resources and λ is the fraction of the resources of any species exposed to predation. These resources are divided between predators such that $\sum_j \gamma_{j,i} = 1$ if species i has any predators, and is zero otherwise. The first two terms in equation 1 are the resource gains of species i from the environment and its prey species respectively, while the final term sums losses to predators. Gains γ are not evenly distributed between predators, hence some predators are more efficient and become more populous than others. Since γ and λ are not in general integer, populations N are not in general integer either.

Unless there is a natural reason to use discretised time, equation 1 is often written as a set of differential equations,

$$\frac{dN_i}{dt} = \gamma_{i,0}R + \sum_j \gamma_{i,j}\lambda N_j - \lambda N_i \sum_j \gamma_{j,i} - N_i, \qquad (2)$$

where the new final term makes explicit the fact that resources are lost between time steps in equation 1. If equation 1 represents discrete generations, resources are lost with the death of the preceding generation. Equation 2 makes such losses an ongoing process, implying overlapping generations.

3.2   Agent-Based Modelling

Agent-based simulation should not need any explanation for this audience, but for the sake of completeness we will explain here how it has been used in the our work.

In agent-based modelling, each individual within the model is tracked separately, so that at any instant it is possible to know the state of any individual, not just the state of the system. Agents also typically have some degree of autonomy, in that they are pro-active software, 'reasoning' about their world and 'choosing' to act, rather than simply being instructed to change state. In an agent-based model, entities of the same type can have some degree of heterogeneity. (For example, in a food web members of the same species may have different size or hunger.) In agent-based social simulation, one is usually concerned with very large numbers of agents, and for the sake of computational time, this limits the complexity of the internal state of these agents (compared, say, to those used by cognitive modellers).

As an example, consider the agent-based model of food web evolution that was constructed to attempt to replicate the results of the system dynamics model based around equation 1. The population dynamics in this model are defined by the interactions of individual agents, following this scheme:

At each time step, the population is shuffled, then for each agent

1. The agent randomly selects a prey, or the world, upon which it will feed, with weighting of prey versus world corresponding to the total population versus number of world resources.
2. If the agent can feed upon its prey (or the world as the case may be), the prey (or a fraction of the world's resources) is consumed, with the agent receiving a proportion, λ of the prey's resources.

3. If the agent received enough resources while feeding (that is, its total resources > 1), it will reproduce. Offspring are given the agent's excess (over 1) resources; the agent is left with 1 resource in its stores.

Note the use of *attempt to replicate* above. As will be discussed in Section 4 in fact it proved impossible to replicate the results. We believe that the failure was due to the use of continuous variables in the equation-based model versus the inherently discrete nature of agent-based modelling – the agent-based model was far more sensitive to low population values than the equation-based model. This was part of the motivation for introducing the third modelling approach: being a discrete equation-based approach, it was hoped that it would help to bridge this gap.

3.3 Individual-Based Modelling

The individual-based simulation is based on the Gillespie algorithm, an essentially exact method for modelling chemical reaction schemes introduced by [17]. This method uses continuous time but discrete states, where individuals are represented by the occupancy of particular states. The model is updated by identifying probabilistically the next event to occur and the time after which it happens based on known transition rates, and updating the states accordingly. Originally envisaged as a means of simulating chemical reactions in cells having only a few molecules of each reactant, a simple scheme might include one simple reactant and its dimer. One possible reaction is that two molecules of the reactant come together to produce a molecule of the dimer. Whenever this reaction occurs, the population of the reactant is decreased by two, and the population of the dimer is increased by one. By incorporating the details of the chemistry into the transition rates and adding more chemical species and transitions, complex schemes can be assembled.

As a worked example of an individual-based simulation, we consider the equivalent reaction scheme to the example in Section 3.1. If there are S species such that $1 \leq i \leq S$ for species i, then we need to track S non-negative integers to monitor the populations. We add a further countable to represent the environmental resources, although these are conserved in the example being considered. Equation 2 can be modelled by three generic 'reactions'. The corresponding reaction rate coefficients are written above the reaction arrow; the total rate of each reaction is found by multiplying this coefficient by the product of the abundances of the reactants. The first reaction,

$$R \xrightarrow{\gamma_{i,0}} R + I, \qquad\qquad [I]$$

models the addition of resources to species i at rate $\gamma_{i,0}R$, with conservation of resources.

$$J \xrightarrow{\gamma_{i,j}} I, \qquad\qquad [II]$$

models the conversion of resources of species j to those of species i at rate $\gamma_{i,j}J$, and

$$I \xrightarrow{d} \emptyset, \qquad\qquad\qquad\qquad\qquad\qquad \text{[III]}$$

models the loss of resources of species i corresponding to the final term in equation 2. We have written the reaction rate coefficient as d although from equation 2, $d = 1$. The total reaction rate of the system is given by summing the individual rates, and the expected time until the next reaction occurs is the reciprocal of this rate, hence

$$t_{\mathrm{exp}} = \left(\sum_i \gamma_{i,0} R + \sum_i \sum_j \gamma_{i,j} N_j + \sum_i d N_i \right)^{-1}. \qquad (3)$$

At this time a reaction is chosen at random with probability proportional to its reaction rate. Supposing that Reaction II were chosen with $i = 2$, $j = 1$, then the population of species j would decrease by one, and the population of species i increase by one.

The advantages of this type of individual based simulation are that some detailed rule-based behaviour can be incorporated at a population level, and membership of relevant groups within a population can be tracked. The large-population behaviour can be approximated with an equation-based scheme, but for small populations the individual-based simulation can produce more realistic results.

4   Finding Common Ground

There was of course some common ground even before the project started; that was the premise for bringing together such a diverse range of groups (which as well as the social simulators and theoretical physicists represented here, included ecologists, geophysicists, biologists and linguists). Despite the diverse problem areas and modelling techniques, many of the system-level properties of the problems studied by the different groups were similar. For example, both the tag-based cooperative groups and the food web showed three recognisable phases: stable relative equilibrium, parasitic behaviour (where one group/species was drawing on resources without supplying any), and collapse of the system. In both cases, the first of these states is obviously the desirable one, so questions arise about how maximise the time spent in that state.

It is not surprising that each group within the project has approached problems using the tools and techniques with which they are familiar. Nor is it particularly surprising that presented with the same problem, different groups would model it in different ways. What we hoped to achieve by looking at the same problems was a better understanding of each others' approaches, in order to derive the general principles that were the main goal of this project.

4.1   Common Ground on Food Web Evolution

The food web evolution problem seemed a natural place for the agent-based modellers to meet the equation-based modellers, as two-species predator-prey relationships are commonly used as introductory assignments for agent-based modelling (often presented as 'wolves and sheep,' or 'foxes and rabbits'), and are also often explored in the context of multi-agent learning (for example, Grefenstette's work[18]). Multi-species food webs would be a natural extension to this. Furthermore, an agent-based approach would allow us to explore some of the assumptions that are encoded in the equation-based approach. In particular, the equation-based approach assumes that individual differences do not impact on the food web structure. For example, in that model one 'resource' is seen to be equivalent to one individual, regardless of the species. In the natural world, members of different species have vastly different sizes. It might be that the larger species have a proportionately smaller number of individuals (which would justify the assumption), but there is no concrete data on this. With an agent-based model we would be able to explore the impact of this and similar assumptions on the results.

However in order to explore the results of loosening any assumptions, it was first necessary to construct an agent-based model that incorporated the same assumptions and replicated the results of the equation-based model. As discussed in Section 3.1, equation-based models are derived from an understanding of individual behaviour, and so it was possible to turn to the original derivation of Eq. 1 and related equations for inspiration on the model. Unfortunately the original formulation of the equations in [9] included one set of equations – those designed to account for competition for resources – that were *not* directly derived from individual behaviour. Without these equations, the model did not generate realistic food webs, so they were an essential element of the model. However because there was no underlying reasoning behind these equations, these equations were hard-coded into the equations, rather than encoding them as "meaningful" agent behaviour.

Having done this, we attempted to replicate the results of the equation-based model. Unfortunately, although we were able to generate some multi-layered food webs, they did not match empirical data (in terms of size and structure) as well as those generated by the equation-based model. In addition, the agent-based model spent far less time in a multi-layered state than the equation-based model, collapsing very easily and frequently to a world with just one or two species. Having said that, the agent-based model was not able to be run over an equivalent time frame as the equation-based model because of the computational resources required. This was a known shortcoming of the agent-based versus equation-based approach, and it is possible that given a longer running time, more stable food webs would have emerged in the agent-based model. Unfortunately we did not have the resources to test this hypothesis.

The simplification of real-valued populations in the equation-based model almost certainly had an impact on the different results in the agent-based model. When species are newly introduced or near extinction, their populations are small, and hence highly susceptible to fluctuations. In particular, it is questionable whether it is reasonable to allow species to persist at populations marginally greater than one, but become extinct if their population becomes marginally less than one. Furthermore, some dynamical systems exhibit cyclical behaviour when a detailed examination using discretised populations is performed, but approach a steady solution if continuous populations are assumed. Dynamical fluctuations in populations could have important consequences for food web structure, and their absence should not be assumed.

It was in part for this reason that the third modelling method was introduced. The individual-based model of food webs captured the system behaviour within a *discretised* set of equations, which still incorporated the assumptions of the original equation-based modelling, but no longer allowed for the existence of fractions of individuals. Statistical properties of the equation-based food web model, such as the number of species and the number of feeding links per species, can be characterised as a function of the resources supplied to the food web [19]. This quantity determines the number of individuals that can coexist. Whereas in the equation-based model species can exist with populations as low as one, population fluctuations mean that the time for which species persist in the individual-based model depends on their typical population, and no definite minimum population exists. The statistical properties of the food web have a similar dependence on resources to that in the equation-based model, subject to a suitable scaling of population. A related consequence of the fluctuations is that the success rate of the introduction of new species is reduced. This, along with the increased computational complexity of the individual-based model, means that as for the agent-based model, only relatively small food webs have been examined, whereas the best-characterised food webs from the equation-based model are the largest.

4.2   Common Ground on Opinion Dynamics

The three problems discussed in Section 2.3 have been generalised to a single abstract opinion-dynamics model. This family of models seems to have some general properties of interest. The dynamics of them seem to fall into two phases: firstly, the opinions roughly converge into groups in $O(n)$ time, reaching a pseudo-stationary state (where n is the number of nodes); then, secondly, the individuals in these groups slowly converge to a single opinion in $O(n^2)$ time. This result seems to be somewhat independent of the topology (it has been proved for a class of these models for all 'non-pathological' topologies in [20]). It is hoped that by relating all these different models as special cases of a single model that a deeper and wider understanding of their properties might result.

This single model has been implemented in such a way as to allow experimentation with different network topologies, and different interaction patterns. It will amongst other things allow us to determine if the analytic results appear to hold for a wider range of configurations. In parallel we hope to explore these problems further from an equation-based perspective. We believe that the equation-based approach will lead us to a better understanding of 'typical' system behaviour, while the agent-based approach will give us insights into 'boundary' behaviour, the unexpected behaviour that may arise in exceptional circumstances (and also an understanding of the exceptional circumstances that may give rise to unexpected behaviour).

5   Conclusions

The attempts to find common ground on this project have highlighted issues in two areas, firstly about modelling in general, and secondly about modelling complex systems. With regards to modelling in general, it has highlighted the difficulties in replicating results using different models. This is not a new finding, and indeed the *Model to Model* workshop series [21] was established for exactly this reason.

In terms of modelling complex systems, our most important result has been the understanding of how the different approaches to modelling complement each other in this area. Equation-based modelling gives useful information about the system as it approaches steady state. It is possible to prove particular properties of the system for particular situations (e.g. particular network topologies) using this approach. Agent-based modelling, on the other hand, allows the exploration of exceptional states, and in particular gives more meaningful results when small numbers are involved. To understand a complex system, it is important to understand both of these cases, as the general case given by the equation-based approach is of course important, but the exceptional states might have critical impact on behaviour, even if they are extremely rare. Once this system understanding has been gained, one can then explore how manipulating the system might alter both the general and exceptional states, and also how it might be possible to control the amount of time spent in particular states.

Acknowledgements

The work in this paper has been undertaken as part of the "NANIA" project, funded by the EPSRC "Novel computation, coping with complexity" initiative. The authors would like to thank all members of the weekly discussion groups who have contributed to the work described in this paper. Additional thanks go to workshop participants for their thoughtful discussions.

References

1. Holland, J.: The effect of labels (tags) on social interactions. Technical Report 93-10-064, Santa Fe Institute (1993)
2. Edmonds, B.: The emergence of symbiotic groups resulting from skill-differentiation and tags. Journal of Artificial Societies and Social Simulation 9(1), 10 (2006)
3. Hales, D.: Choose your tribe! – evolution at the next level in a peer-to-peer network. Technical Report UBLCS-2005-13, University of Bologna, Dept. of Computer Science (2005)
4. Hales, D., Arteconi, S., Babaoğlu, Ö.: Slacer: randomness to cooperation in peer-to-peer networks. In: Proceedings of the Workshop on Stochasticity in Distributed Systems (STODIS 2005) including in the Proceedings of IEEE CollaborateCom Conference, San Jose, CA (2005)
5. Norling, E., Edmonds, B.: Why it is better to be SLAC than smart. In: Proceedings of the World Congress on Social Simulation, Kyoto, Japan (2006)

6. Norling, E.: Contrasting a system dynamics model and an agent-based model of food web evolution. In: Antunes, L., Takadama, K. (eds.) MABS 2006. LNCS, vol. 4442, pp. 57–68. Springer, Heidelberg (2007)
7. Martinez, N.D.: Artifacts or attributes? Effects of resolution on the little rock lake food web. Ecological Monographs 61(4), 367–392 (1991)
8. Dunne, J.A.: The network structure of food webs. In: Pascual, M., Dunne, J.A. (eds.) Ecological Networks: Linking Structure to Dynamics in Food Webs, pp. 27–86. Oxford University Press, Oxford (2005)
9. Caldarelli, G., Higgs, P.G., McKane, A.J.: Modelling coevolution in multispecies communities. Journal of Theoretical Biology 193, 345–358 (1998)
10. Drossel, B., Higgs, P.G., McKane, A.J.: The influence of predator-prey population dynamics on the long-term evolution of food web structure. Journal of Theoretical Biology 208, 91–107 (2001)
11. Drossel, B., McKane, A.J., Quince, C.: The impact of nonlinear functional responses on the long-term evolution of food web structure. Journal of Theoretical Biology (229), 539–548 (2004)
12. Quince, C., Higgs, P.G., McKane, A.J.: Deleting species from model food webs. Oikos (110), 283–296 (1995)
13. Neau, D., Amblard, F., Weisbuch, G., Deffuant, G.: Mixing beliefs among interacting agents. Advances in Complex Systems 3, 87–98 (2000)
14. Baxter, G., Blythe, R., Croft, W., McKane, A.: Modeling language change: An evaluation of trudgill's theory of the emergence of new zealand english (in submission) (2008), http://www.ph.ed.ac.uk/~rblythe2/Preprints/BBCM08.pdf
15. Baxter, G.J., Blythe, R.A., Croft, W., McKane, A.J.: Utterance selection model of language change. Physical Review E 73 (2006)
16. Edmonds, B.: Achieving consensus among agents - an opinion-dynamics model. Technical Report CPM-08-185, Centre for Policy Modelling (2008)
17. Gillespie, D.T.: Exact stochastic simulation of coupled chemical reactions. Journal of Physical Chemistry 81(25), 2340–2361 (1977)
18. Grefenstette, J.J.: The evolution of strategies for multi-agent environments. Adaptive Behavior 1, 65–89 (1992)
19. Powell, C.R., McKane, A.J.: Effects of food web construction by evolution or immigration. Submitted to Oikos (2008)
20. Baxter, G.J., Blythe, R.A., McKane, A.J.: Fixation and consensus times on a network: a unified approach (in submission) (preprint) (2008), http://arxiv.org/abs/0801.3083
21. M2M 2007: Third international model-to-model workshop (2007), http://m2m2007.macaulay.ac.uk/

Towards a New Approach in Social Simulations: Meta-language

Raif Serkan Albayrak[1] and Ahmet K. Süerdem[2]

[1] Yaşar University, Izmir, Turkey
raif.albayrak@yasar.edu.tr
[2] Istanbul Bilgi University, Turkey
asuerdem@bilgi.edu.tr

Abstract. In this paper we will present a framework for bridging micro to macro emergence, macro-to-micro social causation, and the dialectic between emergence and social causation. We undertake a cultural approach for modeling communication and symbolic interaction between agents as the key element of connecting these three aspects. A cultural approach entails modeling cognitive agents who are not only capable of representing knowledge but also able to generate meanings through their experiential activities. We offer a meta-language approach allowing dynamic meaning generation during the interactions of the agents. This framework is implemented to a social simulation model. There are four important implications of the model: First, model shows a dynamic setup where agents can generate and elaborate multiplicity of meanings. Second, it exemplifies how individual mental models can interact with each other and evolve. Third, we see that a thickly coherent cultural background is not necessary for the emergence of embedded social networks, a thin coherence such as opposition maps would be sufficient to observe their dynamic formation. Fourth, exchange of meanings through successful sense-making practices generates a social anchoring process.

Keywords: Social simulation; emergence; culture; language; reflexivity; intentionality; semiotic relations; meaning generation.

1 Introduction

Agent Based Social Simulation approaches typically assume that emergent social patterns can be reduced to the properties of their components. They start by modeling built-in cognitive devices to individual agents; determining the rules of interaction among them; and hence trying to observe emergent macro patterns that influence the behavior and interactions of the individual actors [12]. However, this is only possible if it can be assumed that micro-level behavior could be functionally determined. For social systems such functionality is bestowed by rules and habits [27]; shared social norms [17] and goals and values setting social institutions [38]. Norms and rules make feed-back between the interactants predictable and thus emergence of social order possible.

N. David and J.S. Sichmann (Eds.): MABS 2008, LNAI 5269, pp. 195–214, 2009.

These premises have their origins in the bounded rationality theory [36]; evolutionary economics [2] and methodological individualism [27]. Main arguments can be summarized as follows: What makes human beings a social animal is not inscribed in their biological genes but is imminent in the cultural codes. Individuals internalize cultural codes as a second nature through their capacity for behavioral learning. Within this framework, rules and norms are recorded in the memory as patterned behavioral responses to stimuli; in other words they are stored as built-in cognitive procedures. Hence, habits of association and social learning embody social norms to induce characteristics akin to personality traits within the human psyche. These codes hold a society together since they determine the rules of the game, shared values for the actors with different goals and preferences.

One of the widely accepted formulations of this approach is by Geert Hofstede who defines culture as "...the collective programming of the human mind that distinguishes the members of one human group from those of another. Culture in this sense is a system of collectively held values" [28, p. 5]. Hofstede receives culture as constitution of orthogonal value dimensions existing as latent variables which manifest themselves through a multitude of directly observable indicators. These latent variables are operationalized through a dimension reduction technique gathering several indicators to construct a composite index for each value dimension. These composite indices are then used for measuring cultural differences among different societies. A multi-agent simulation application of this approach is present in this volume which takes Hofstede's power distance dimension to formulate behavioral rules for artificial trading agents [29].

However, this approach is problematic for understanding overall behavior of complex systems where the interaction between the system and the individuals involve causal feedback loops. In case of complexity, the system cannot be reduced to its parts and information from emerging macro institutionalized practices becomes an independent variable itself [41]. Agents are autonomous and heterogeneous at the micro-level and are influenced by their conception of the emerging macro patterns of behavior. The chain of reference and causation in a social system is at least partially self-referential when individuals base their decisions on others' actions and utterances [18]. Interaction between agents becomes fuzzily reciprocal and this makes it difficult to determine any reducible and coherent rule. Culture as latent variable approach, on the other hand, assumes that agents should be coherently interconnected in terms of a structured value system. Yet, in case of complexity the interconnection between the individuals and the structure is in constant flux hence interaction cannot be lawfully related. This automatically invalidates the assumption that a collective programming of mind in terms of shared values connects the individuals between themselves and with the system.

Capturing the complexity of emergent macro behavior patterns requires a culture theory which is able to go beyond its definition as a structured system of shared values or a portfolio of predefined normative protocols. Sawyer [40] proposes connecting emergence theories to the study of symbolic interaction for solving the problem. As the heterogeneity of the rules of symbolic interaction between agents increases, the more emergent structure becomes irreducible

[40], [18]. Therefore, modeling culture as symbolic interaction and endowing the agents with the capacity to re-cognize emergent macro behavior becomes essential for a sound social simulation model. Such an endeavor will provide the modelers to connect three aspects of the micro-macro link in sociological theory: micro-to macro emergence, macro-to-micro social causation, and the dialectic between emergence and social causation. One of the remarkable efforts to bridge the micro-macro link is provided by Dignum et al. [14] in this volume, offering a mediating meso-layer connecting the macro and micro level in a simulation. The meso layer covers the elements influencing the behavior at the group level. They define these elements as norms and organizational (or group) structure. Different from the culture as a latent variable approach, these elements are not formulated as cultural scripts dictating behavior at the micro level and different form the top down approaches they are not causal laws used on the macro level.

In this paper we will attempt a similar effort for bridging micro to macro emergence, macro-to-micro social causation, and the dialectic between emergence and social causation. The intrinsic difficulty of modeling causal feedbacks to the lower levels is the problem of imposing structures to emerging macro patterns by the designer; the artificial society should be self-amending. We take communication and symbolic interaction between agents as the key element of connecting the three aspects of the micro-macro link and as a solution to this problem. Our view of communication goes beyond its definition as codification of symbolic interaction used as a medium of information transmission. Building agents transmitting, receiving and processing information and accordingly give behavioral responses will not be sufficient to model symbolic interaction as a form of emergent macro phenomenon since units of information in macro and macro levels are essentially different. Such a model requires a cultural approach endowing the agents not only with the capacity to recognize but also to make sense of the information from the emerging group structures at the meso level. In this respect, we share the same problematic with Dignum et. al. On the other hand, unlike their model, our model does not separate the cultural and cognitive aspects of behavior; it allows for negotiating for the emerging group level (meso) network structure.

1.1   Modeling Cognitive Agents

A cultural approach entails modeling cognitive agents who are not only capable of representing knowledge but also able to generate meanings through their experiential activities. For that, we need to go beyond models of symbol manipulation and model symbolic interaction as embodied and motivated by everyday practices. We postulate that linguistic ability is more than manipulation of abstract symbols; its use draws upon all cognitive resources [13]. We do not limit presentation of cognitive system to a theory of functioning of cerebral activities at an abstract level but also take into account the sensory processes and the body-world interaction. Human intelligence is a dynamic and complex process formed by the body, the world and the brain. Moreover, human cognition is not a simple reaction to signals from the external environment. Repetitive interactions between the organism and the environment constrain possible responses by the organism and thereby constitute the emerging cognitive processes. Mind is embedded in and embodied by the emergent patterns constituted by these interactions.

Understanding mental processes thus requires considering the organism and the system as a whole and the "sensual" (connotational) character of perception. Peculiarities of the emerging perceptual patterns (e.g. Gestalt) constitute our conceptual structure and linguistic practices [33]. The theory of embodied cognition is recognized by the AI community and there are efforts to redefine cognition and behavior in terms of dynamic interaction between brain, body, and environment [3]. The basic premises of this endeavor is grounded in the ideas of Merleau-Ponty [37] who claimed in his Phenomenology of Perception that consciousness, the outer world and the body are reciprocally intertwined during the perception process. Embodied cognition theorists recognized Merleau-Ponty's legacy and widely used his phenomenology for criticizing the traditional AI approaches, e.g. [7] and [16]. According to Merleau-Ponty the phenomenal is not an object out there but is constructed through our bodily and sensory functions. In other words, intentional objects of thought (noumenal) cannot be separated from the perceived objects of thought (phenomenal).

Cognition is embodied and this embodiment resides in the unique history or "the lived" experiences of the body (le vecu). However, people do not live in a social vacuum and social interaction is the essential lived experience mediated through culture. The unique history emerges as a pattern of shared experiences through structural couplings. When two or more people interact, their lived experiences mutually modify each other and their mental system embodies the perturbations created by the emerging interactions. When these interactions start to constitute repetitive patterns, individuals become "structurally coupled" at the group level hence become embedded in the emerging structure [24]. Once structurally coupled, shared experiences start to become the common "reality" or "life world" [37] of the interacting individuals.

Emergent life-world constitutes a base for social coordination and emotional and cognitive codes created by the common experiences orient individuals to construct identities, coordinate action, and create cooperation at the group level [42]. In that respect, life-world is the common leitmotive underlying capabilities, practices and behaviors residing cognitive repertoires of the individuals that form a community [25]. On the other hand, structurally coupled life-world cannot be reduced to static recurrent patterns that enact the same structure permanently at the group level. The emergent life-world provides the individuals with a habitus [6], with fuzzy rules like feeling of the game or practical sense allowing them to generate infinitely many situation dependent strategies with their available cultural resources. Bourdieu's definition of habitus extends structural coupling of shared experiences beyond direct interactions between individuals at the group level to the macro structuring of the social space. That is, geographically distant individuals may share the same habitus if they are proximate within the social and symbolic space and experience similar practices in their everyday life. In that sense, habitus mediates between the macro structure, emergent group behavior and individual "lived experiences" of social agents within a shared symbol space.

1.2 Symbol Systems and Decoding Emergent Structures: Reflexive Agents

So far we have elaborated social theories explaining how social interactions and everyday practices become embedded in emergent patterns of interaction. We have not yet referred to how social actors decode and feedback knowledge from macro structures. According Goldspink and Kay [24] human capacity for language is the key to model fuzzy and complex relations between the components of a social system. Contrary to natural systems where the behavior of the individuals are activated by local influences only, social systems can handle the problem of complexity through a feedback mechanism which allows changes induced in the macrostructures to be felt locally. This feedback mechanism occurs by means of linguistic activities that provide agents with reflexive capacities. Only reflexive agents would be able to decode macro-patterns and encode their local behavior. They would have the capacity to distinguish emerging phenomena at multiple levels and language provides them with a foundation for flexible and instant feedback about the emergent structures.

Gilbert [22] modeled agents that can detect the presence of emergent properties and communicate them with a capacity for symbolic interaction. His agents were endowed with binary valued "tags" which could be interpreted in terms of a variety of attributes such as gender, race, and so on. Instead of being built-in properties, these tags enabled agents to act individually and perceive the tags of other agents. Subtlety of Gilbert's agents lies in their ability to decide on which one of the tags would be their significant characteristic instead of being entities with a priori properties defined by the modeler.

However, we need more to model symbolic reflexivity than recognizing and enacting emergent tags. Linguistic abilities encompass not only recognizing and describing features but also expressing deliberate relations [34]. In other words, reflexive action is not limited to the symbolic affirmation of shared social classifications and standard rules that regulate social relations but involves symbolic construction of intentionality such as motives, preferences and goals. Building reflexivity into emergence therefore requires modeling deliberate agents whose behavior is regulated by discursive anticipations of future group structure besides their capability of using or selecting existing features, properties or rules [9]. Emergence is future based and therefore depends on the negotiating acts of the agents. Thus, building deliberate agents requires enabling agents with the faculty of recognizing and reflecting on the environment-self relationship; or in other words building "consciousness" to agents. Castelfranchi [9] proposed to reconcile cognition and emergence, intentionality and functionality by modeling and simulating agents that can have the consciousness (our term) of the emergent phenomena and use the cognized information as a feedback to reform their minds and reproduce themselves. Modeling such agents needs to go beyond implementing associative learning based on taxonomies, rules, associations etc. but should operate on cognitive operations such as beliefs and goals.

1.3   Modeling Reflexivity

Thus, building agents with reflexive capacities requires endowing them with a certain ability to generate, objectify and institutionalize knowledge as "common sense" reality, a factual belief system and communicative mechanisms of creative adoption of knowledge patterns taken from the collective "stock of knowledge" [42]. Modeling belief systems is important if we want to design "cognitive agents" because agents' goal adoption; preferences; decisions and actions are based on their beliefs [9]. Beliefs on the other hand do not hang on by themselves but assemble into coherent systems of ideas based on a few basic axioms about the world. Factual belief systems, endow the individuals with strategic intellectual tools for categorizing the world.

As theory of embodied cognition states, cognition, everyday practices and language are embedded in each other. Sapir-Whorf hypothesis assumes that semantic-syntactic organization of a language is systematically related to the belief systems organized as the worldview of a people through the construction of the causality mechanisms of perceiving the world and the conceptual classification of entities [35], [10]. This relationship provides individuals with a priori thoughts and intuitions for organizing separate impressions of the world. Since reality is complex, conception of phenomena results when the individual interprets them by relying on intuitive causal relation schemas. Personal theories of different individuals lead them to conceive the same noumenal reality as phenomenal instances of their personalized concepts [32].

Modeling factual belief systems in terms of designing mental models that serve as internal representations of the world is not new to social sciences. One of the well established of these devices is the "mental mapping" technique. This technique is based on the premises that mental models are internalized representations of the world grounded in linguistic accounts which can be elicited as networks of concepts. Within this network, a concept makes sense in relation to other concepts for an individual and the social meaning of the concepts are not universal but constituted as the intersection of the individuals' mental models [8]. Mental mapping technique depends on a sound formalization ground within the mathematical graph theory by representing the concepts as vertices and the relations as edges where a pair of vertices linked by an edge is referred as a statement and the whole text corpus as a map. In this volume, Tsvetovat et al. [46] provide us with an example of modeling belief structures of agents as a form of concept network used for storing acquired interaction histories as edge properties and encode the agents' transactive memories. In our perspective, mental models serve to interconnect the agents with a habitus like structural coupling where individual agents need not to be direct neighbors in order to interact.

However, mental mapping techniques are limited in terms of endowing agents with reflexive capabilities besides their capacity to model the factual understanding of the world. Their design follows an extensional logic requiring instant resolution of concept-meaning-practice correspondences. In logic, extension represents the denotative element during a reasoning process since it equalizes the concepts related to the same set of objective properties according to a predicate, while intension is connotative since it refers to the set of all possible things a word or phrase could describe [5]. Mental maps are denotative; they possess a strongly coherent character, leaving no space for interpretations and/or negotiations for the counterparts in a cultural interaction.

On the other hand, modeling reflexive agents during cultural interactions entails a connotational element. The agents must be allowed for the capacity of a multiplicity of interpretations over concepts although they might have a temporarily preferred one that can be properly represented as a mental map. Even in their dynamic form (see [46] in this volume), mental maps won't help the agents to interpret the intents of their counterparts and their messages for future cooperation; cooperation is taken for granted for agents with similar mental maps. Everyday practices are full of instances that put discursive act assumptions at stake and reveal a plethora of interpretations. Hence, interpreting the intents of the counterparts and imagining their effects on the emerging cooperative act is the essential aspect of symbolic interaction.

Knowledge is socially constructed as an emerging process embedded in the already accumulated collective stocks of knowledge and is mediated by the cultural devices such as symbol systems rather than being built-in, already in the mind constructs [4]. Cultural devices provide embedded individual actors with a historical a priori. Geertz [20] points to the public nature of the symbol systems during his criticism of cognitive anthropological theory which has been a theoretical base for mental modeling approaches. Culture consists of socially established structures of meaning within which people negotiate the intents of their actions. Geertz agrees with cognitive anthropologists that rather than rational motivations and interests meaning should be the object of human studies. Yet, he emphasizes that meaning generation requires a high degree of mutual understanding making it a public rather than a private affair. He contests the definition of culture as a mental model providing the individuals with what they have to know or believe in order to operate in a manner acceptable to the other members of a society.

Although Geertzian culture theory provides us with a public frame for modelling the negotiation of meanings of individual intentions, it carries a deficit when it comes to studying causal relationship between culture and social action. Conceiving culture as a thickly coherent, static, and singular form that can only be understood by referring to itself, this theory not only turns the fundamental relationship explored in mental map approaches (culture to action) upside down but also dismisses the expected functionality of this relationship (causality). We will end up in an infinite regression within the symbol space if we were to apply Geertzian framework for modelling communication between the agents. In a simulation model that intends to use such a model, there could be a single objective for agents: solving the "symbol grounding problem" [26] or in other words to learn Chinese from a Chinese to Chinese dictionary [43]. Thus, without an anchor, attempting to solve/model a purely self-referencing problem is a futile effort. We believe that such an anchor can be found in Sewell's [44] culture theory which successfully combines the definition of culture as an embedded activity in everyday practices and to its definition as a symbol system within a "thin" semiotic approach. In the rest of the paper, we will present and operationalize this theory for modelling reflexive agents capable of negotiating and constructing meanings. We modify mental map approach as "opposition map" to allow for a "thin" semiotic coherence; and propose a simulation framework modelling a dynamic semiotic web which feed-backs from individual practices of the agents.

2 Formalization of the Cultural Framework

2.1 Opposition Map

Sewell's [44] culture theory first focuses on the subjectivity of meanings. Actors are allowed to construct and reconstruct meanings for a set of symbols taken from a publicly available symbol system and deploy them strategically to pursue their own goals. On the other hand, this subjectivity is limited since developing a strategy is contingent upon the available set of cultural resources. Recursively, as particular cultural resources become more central, "and become more fully invested with meaning" [45, p. 281], they anchor the strategies of action people have developed. Then culture is not only the set of constraints that restricts the human behaviour, but also is the mediator of interactions that can be bended, kneaded and put into many forms.

Sewell completes "culture as practice" approach with the "culture as a symbol system" approach which takes culture as a network of semiotic relations pervasive across the society; a network that constitutes the interaction between symbols, meanings and practices. The basic element of this network is symbols and in order to realize their existence they must be distinguishable. In a roughly Saussurian way, we may argue that every symbol reveals its identity from its distinctions to other elements in the same system. Saussure [39] labeled these distinctions as oppositions. Oppositions are the basic operations for meaning generation; "the binary opposition is a child's first logical operation" [30, p. 60]. Thus, to make sense of the symbols individuals need to share the same logic of oppositions. As Sewell states: "the users of culture will form a semiotic community in the sense that they will recognize the same set of oppositions and therefore be capable of engaging in mutually meaningful symbolic action" [44, p. 166]. This statement forms the base of the opposition map as an alternative to mental map approach and in the following lines we will formalize it.

Since opposition relation is binary and is congruent to the distinguishing character of symbols stressed above, we infer the opposition among experiences, or practices or meanings by defining an algebra over opposition relations among symbols. This scheme not only formalizes Sewell's premises but also extends it by defining the fundamentals of an appropriate dynamic framework. To us it is quite surprising that literature lacks a formal algebra of oppositions despite the popularity of Saussurian semiotic relations (but see [19]). Therefore we start by developing an algebra of binary oppositions.

We assume the existence of a symbol set $S = \{s_1, \ldots, s_p\}$ that consists of a finite number of symbols. An opposition $\mathcal{O}$, is a binary relation defined over $S \times S$ and satisfies,

1. For all $s_m \in S$, $(s_m, s_m) \notin \mathcal{O}$. That is, the relation is irreflexive.
2. For any $s_m, s_n \in S$, if $(s_m, s_n) \in \mathcal{O}$ then $(s_n, s_m) \in \mathcal{O}$. That is, the relation is symmetric.

First condition states that a symbol can not oppose itself since opposition relation is instrumental to guarantee the existence of that symbol among the others. Therefore the relation must be irreflexive. Second, we argue that if one can distinguish a symbol in the existence of the other, then the latter must be distinguishable in the existence of the former. This is a consistency principle which assumes the existence of cognitive abilities that goes beyond simple book-keeping of one way opposition relationships. Whenever a symbol enters the system, opposing to some particular symbols, system responds this perturbation as a whole and locates the incoming symbol to a corresponding setting. This reflex requires symmetricity. It also explains how the system handles granularity. In other words, if a symbol is distinguishable within the existence of a set of symbols but within no proper subset of it, due to symmetricity, this set of symbols enters the system as a symbol of its own (neither gin nor tonic, but as gin-tonic) and does not necessarily preserve the oppositions of its constituents. One further note is that opposition relation is not necessarily transitive. Hence if s_1 opposes s_2, and s_2 opposes s_3, then it is not necessarily the case that s_1 opposes s_3.

These properties delineate the structure of the mental map in our model. Symmetricity condition allows us to represent the binary opposition relations as bi-directional networks while irreflexivity makes sure that there are no self loops. We call these networks as opposition maps. Opposition maps are not abstract constructs, we can empirically elicit them from narrative data using a semantic mapping perspective explained in [1].

Fig. 1 illustrates the opposition map of a political actor elicited from a set of policy discourses; namely social security reform in Turkey. Data includes in-depth interviews, public declarations and reports of the agents on the issue.

Opposition map locates symbols with respect to each other in such a way that it constitutes a loose mental schema. The entire thrust of this schema for our purposes is to implement a deconstructionist perspective that reveals the instability of linguistic meaning. According to Sewell, "the meaning of a text or utterance can never be fixed; attempts to secure meaning can only defer, never exclude, a plethora of alternative or opposed interpretations" [44, p. 167]. While meanings of symbols are floating in public, opposition maps are individual mental schemas. Therefore they are not subject to continuous evaluation, but perturb only in a radically discrete manner, in the form of a cultural shock [45], [15], [44].

2.2   Meaning Generation

Thus, opposition maps let us to model mental schemas while allowing an intensional logic open to reflexive negotiation. Yet, we need to go further from

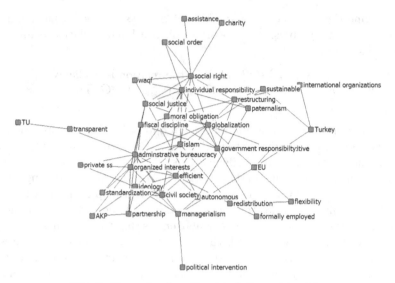

Fig. 1. Example of an Empirical Opposition Map

opposition maps in order to obtain a dynamic model that can be used to simulate meaning generation. We extend the algebra of binary oppositions to a meta-language frame which connects individual opposition maps to each other in a dynamic manner and generates meaning. Let's start with the definition of meaning. Technically, meaning of a symbol is defined from the oppositions of a symbol to other symbols in the same system. There are two basic premises for this statement: First, a meaning is merely a set of symbols and second, there are no opposing symbols in that meaning set.

To formalize, let's first define opposition class of a symbol s_n, as the set $\bar{s}_n = \{s_m | (s_m, s_n) \in \mathcal{O}\}$. Thus, a meaning, m, is a subset of S such that if $s_m \in m$, then $\bar{s}_n \cap m = \emptyset$.

In the simple opposition map example illustrated in Fig. 2, the set $\{s_2, s_4, s_5\}$ constitute a meaning since it contains no binary opposing symbols. Similarly,

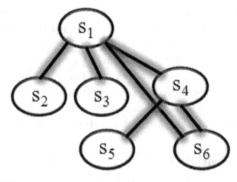

Fig. 2. Example of a Simple Opposition Map

sets $\{s_2, s_6\}$, $\{s_2, s_4, s_6\}$ and $\{s_4, s_5, s_6\}$ are all meaning sets among many others that can be revealed from the opposition map. However the formal definition of a meaning does not specify the symbol it explains because it is purely symmetric in terms of its constituents. In this regard we assume that individuals can generate meanings and whenever a meaning is generated, it is actually assigned to all constituting symbols. The principle endeavor of our framework is to formalize all possible interpretations of symbols. This definition of a meaning set implies that given an opposition map one can reveal numerous meanings of a symbol. For example, meanings, $\{s_2, s_6\}$, $\{s_2, s_4, s_6\}$ and $\{s_4, s_5, s_6\}$ are all valid interpretations for s_6 with respect to the opposition map displayed in Fig. 2. Although multiplicity of interpretations regarding a symbol allows being reflexive and negotiating meaning, the cost of assigning meaning to all alternatives may be higher than the benefits. In real life when this definition is employed to capture the interpretations of a concept, one may end up with an exclusive list of rubbish entries, since a list of mutually non-opposing symbols would not necessarily affirm each other and have no practical significance. However this can be optimized by grounding meanings in practices and we explain this by a simple example demonstrating also a simple meaning generation exercise.

Meaning Generation: Example

In this example there are two agents, a man and a woman belonging to the same primitive tribe. They are identical in a sector of their opposition maps but they generate different meanings regarding the symbol *rain* as displayed in Fig. 3. When these agents observe or negotiate *rain* symbol, they interpret it according to their corresponding meanings. For instance, man uses a single meaning, {rain, dark, cold} and makes sense of the *rain* within the context defined by *dark* and *cold*. As long as nature proves otherwise agent uses this meaning as his personal theory. Maybe his primary concern is hunting and because of *rain* it would be *dark* thus the prey would not notice him while he will get wet and feel *cold*. The symbols prey and wet are not displayed in this segment of the opposition map. On the other hand, for the woman the situation is different because she has two meanings to interpret *rain*: {*rain, cold, moon*} and {*rain, dark*}. With respect to the first meaning set maybe she realizes that *sun* disappears and *moon* replaces its exact spot and it becomes *cold*er. With respect to the second set, when it *rain*s it gets *dark*er, maybe it would be better to call children back to cave. We see two roles that the woman plays here; an ambitious observer and a mother. As an observer she explores but as a mother she acts pragmatically.

This accounts for the optimization that we have discussed previously. She needs to interpret the observed symbol therefore she must get rid of or at least decrease the complexity by devising an ordering mechanism for the corresponding meanings. This ordering is contingent to her practices with the nature. Here complexity is defined as a one-to-many assignment of symbols and meanings. As the agent makes sense of the situations of observing (*rain*) repetitively, depending on the remuneration she receives, a relative ordering over meanings is constructed and complexity is gradually reduced. Here, lets assume that after

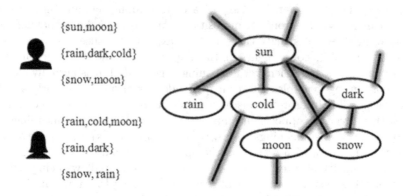

{sun,moon}

{rain,dark,cold}

{snow,moon}

{rain,cold,moon}

{rain,dark}

{snow, rain}

Fig. 3. Opposition maps of a man and a woman belonging to the same tribe partially coincides with each other. However they have generated different meaning for the symbols.

some time our primitive woman stopped using the set {*rain, cold, moon*} since it does not make any practical sense to her (maybe being an explorer does not pay) and kept the second one for her children's welfare. However, inverse scenario is equally probable. If the collection of meanings (*rain*) is incapable of making sense then the agent infers new meanings from her opposition map. The dialectics between the urge to make sense of situations and the urge to reduce uncertainty shapes and reshapes meanings assigned to symbols.

2.3 Meta-language

Up to now we have demonstrated how reproduction capacity of the meaning system is embedded in non-oppositions that are fine tuned through practices with a simple example. Next step is to develop an algorithm that is able to generate meaning sets from opposition maps. Since a meaning can be conceived as a list of symbols to which it is associated, the collection of all meanings that an agent reveals from his/her opposition map efficiently describes all symbol-meaning correspondences. Furthermore if this collection covers the symbol space completely, then it is the personal theories of an agent about everything, his/her world view. We label this collection as *meta-language* and formalize it.

We start by the usual definition of a cover of a set. A collection of sets $X = \{x_1, \ldots, x_g\}$, where each $x_i \subseteq S$ is said to be a cover of S, if for all elements $s \in S$, there exist at least one $x_i \in X$ such that $s \in x_i$. A meta-language $M = \{m_1, \ldots, m_k\}$ is a set of meanings for the opposition map $(S, \mathcal{O})$ that covers S.

Meaning generation is an irreducible task since it is crucial to conform to the opposition map as a whole. As the number of symbols in opposition maps get larger, it becomes harder to check for the consistency argument that no opposing symbols belongs to the meaning set. This is a problem that an agent faces during complexity reduction (i.e. to relax one-to-many assignments of symbols to meanings). Merged meaning sets come into existence as a consequence of the trade-off between sense making and complexity reduction. The problem here is the increased difficulty in handling the construction of these sets. Situation gets worse for the generation of a meta-language that consists of many such meanings.

We now state and prove a theorem that explains how minimal amount of cognitive capacity would be enough to generate a meta-language in a huge symbolic space such as the social space. The principle idea is developing base sets by which all meta-languages can be constructed, more or less in the same manner one can construct any vector in a vector space by using unit vectors. The base sets used to construct meta-languages is called meta-language generating set.

Given an opposition relation $\mathcal{O}$, over a set of symbols S, a meta-language generating set, $\mathfrak{M} = \{\mathbf{m}_1, \mathbf{m}_2, \ldots, \mathbf{m}_k\}$ consists of subsets $\mathbf{m}_i \subseteq S$ such that for any meta-language M of $(S, \mathcal{O})$, and for any $m_i \in M$, there exist at least one $\mathbf{m} \in \mathfrak{M}$ such that $m_i \subseteq \mathbf{m}$ and for any $p \neq q$, $\mathbf{m}_p$ is not a proper subset of $\mathbf{m}_q$.

Like all meta-languages the elements of meta-language generating set $\mathfrak{M}$ are sets of symbols. Also if M is an arbitrary meta-language and for any $m_i \in M$, m_i cannot contain any $\mathbf{m}_i \in \mathfrak{M}$. The following theorem not only guarantees the existence of a meta-language generating set for any symbol set and binary opposition relation pair but also proves that this set is unique.

Theorem 1. Any opposition relation $\mathcal{O}$ over a set of symbols S defines a unique meta-language generating set $\mathfrak{M}$.

Proof. Proof constructs set $\mathfrak{M}$ and shows that it is unique. Let $\bar{\mathcal{O}}$ define the complement of the binary opposition relation $\mathcal{O}$, $\bar{\mathcal{O}} = \{(s_i, s_j) | (s_i, s_j) \notin \mathcal{O}\}$. Let $\mathfrak{G}$ be the undirected graph representation of $\bar{\mathcal{O}}$.

Proof continues in graph theory framework. Construct $\mathfrak{M}$ such that $\mathbf{m} \in \mathfrak{M}$ if and only if $\mathbf{m}$ defines a set of nodes that forms a clique in $\mathfrak{G}$. A clique C is a subgraph such that each node is connected to every other node and the set is maximal with respect to this property. Clearly, there may be more than one clique within a graph.

Thus each element of $\mathfrak{M}$ contains cliques as sets of nodes. Since the set of cliques of a graph is unique and is a cover, $\mathfrak{M}$ is unique and is a cover of S. From the construction of $\mathfrak{G}$, $\mathfrak{M}$ is a meta-language.

It only remains to show that if M is any meta-language of $(S, \mathcal{O})$, and $m_i \in M$ then $m_i \subseteq \mathbf{m}$ for some $\mathbf{m} \in \mathfrak{M}$. In other words any meta-language contains sets that are subsets of some elements of $\mathfrak{M}$. Since there can be no opposing pairs within m_i, it follows that a non-opposing graph representation of m_i is a complete sub-graph of $\mathfrak{G}$. If m_i is not contained properly in another complete sub-graph then m_i is a clique so, $m_i \in \mathfrak{M}$. On the other hand, if m_i is not a clique then it must be contained in some clique $\mathbf{m} \in \mathfrak{M}$ such that $m_i \subseteq \mathbf{m}$. This completes the proof.

According to the theorem, meta-language generation runs over complete sub-graphs of non-opposing symbols. In other words, in a world populated with lots of symbols, locally non-opposing symbol domains would be sufficient to generate meanings and hence meta-languages.

Up to now we have formalized the fundamentals of a meta-language which generates all possible meaning sets of an individual. The entire thrust of this formalization is to ground this perspective to a working model of sociality. Since our knowledge about practices, desires, and intents is socially constructed through a discursive process, such individual motivations are not nomothetical entities but negotiated meaning domains. A social construct is a naturalized concept or practice in the eyes of those who give a common sense meaning to a shared semiotic symbol. Individuals and groups participate in the creation of their perceived social reality [4]. Within this framework, objects of all practices can be reduced to symbols and their meanings. In the next section we propose a simulation based implementation of meta-language where the agents interact to make a synthesis of their personal theories with their social practices on a reflexive basis. This implementation will provide us with a model demonstrating how agents discard the meaning domains which makes no practical sense to them through symbolic interaction.

3   Implementation of Meta-language

Our example in the previous section provided us with a simple model of how an individual makes sense his/her individual practices when interacting with nature. Social interactions, on the other hand, are different since they involve two reflexive parties and reflexivity happens in and through meaning exchange [16]. In our model, meta-language allows agents to play on the multiple meanings of symbols, in such a manner that agents may redefine situations in ways that they believe to favor their intents. This opens the door for reflexivity. Incentives, evaluation criteria and intent of each agent may be defined by the context of other spheres of social life such as economics or politics. However, semiotic structure is autonomous from these spheres. Although an action may be motivated to some extent by other spheres, it still needs to be decoded as a meaning in meta-language to make sense. As we have demonstrated, all socially constructed objects can be reduced to symbols and their meanings.

Thus, social interactions do not occur in a semiotic vacuum: When people interact, they do so with the understanding that their respective personal theories are related, and as they act upon this understanding, their common knowledge of reality becomes reinforced. As agents develop their own metalanguage models, social interactions loads inherited information to a code. People sharing same meaning system forms a semiotic community.

According to Sewell, people who are members of a semiotic community are capable not only of recognizing statements made in a semiotic code but of using the code as well; of putting it into practice. To use a code means to attach abstractly available symbols to concrete things or circumstances and thereby to posit something about them [44]. Moreover, to be able to use a code means more than being able to apply it mechanically in stereotyped situations it also means having the ability to elaborate it, to modify or adapt its rules to novel circumstances [44]. Hence, the system is always rich enough to generate new theories, new meanings to make sense of the symbols. As new meanings are generated and publicized, some become popular and are shared across the members of a community while others diminish. Meaning generation is an individual reflexive action, on the other hand exchange of meanings through successful sense-making practices is a social anchoring process.

When an individual reflects upon a meaning domain in the light of practical judgment, that individual accumulates habitus like personal or tacit theories. These tacit theories can be acquired through a reflexive process where redundant meaning domains are discarded according to practical judgment. We now model how practical judgment selects symbols through social exchange by nominating meanings to them.

3.1   A Sense Making Multi Agent Simulation Set-Up

This set-up demonstrates an implementation of the cultural framework discussed in a model where the interactions between agents are purely symbolic as a sense-making game. There may be various applications: to capture social network formations; analyze the dynamics and/or the stability of formed groups as a result of symbolic interaction and so on. In this example, we particularly aim to relate embeddedness of social transactions in terms of triads in a social network [11] to the degree of semiotic coherence.

Simulation sets off with populating the symbol set with nine symbols. Agents are distributed in a 10 by 10 grid (initially 80% full) with opposition maps randomized with respect to a pre-specified criterion such as a 70% density. For all agents meta-languages are randomly assigned according to the algorithm in Theorem 1. Since there can at most be $2^9 = 512$ different meaning sets, all meaning sets are labeled with reference to the exhaustive list. Fig. 4a displays a meta-language assignment for an agent. In this assignment each s_i is a symbol and each m_j is a meaning set composed of symbols. Clearly the assignment is a correspondence. Furthermore for each meaning, a random reliability is assigned (Fig. 4b).

All agents in the grid have random probabilities to interact with each other that is offset with the inverse square law of distance in the grid [23]. Agents reproduce with a fixed probability to give birth to an offspring whose opposition map mutates with a constant rate. Grid is pruned from randomly occurring deaths where the rate of survival of an agent depends on the number of successful interactions.

In each turn in the simulation, every agent plays an active role to initiate a social interaction. Active agents select passive agents randomly. Interaction starts with a signal from the active agent to the passive agent in the form of a meaning set representing his intentions. The meaning that the active agent selects is randomized with respect to reliabilities. Passive agent evaluates the meaning set whether it belongs to a clique in his meta-language generator. In other words, tests the conformity of the meaning with his opposition map.

Our framework suggests symbol-meaning correspondence to be dynamic for any agent. A symbol might not only correspond to multiple meanings but also

a.
$$\mathfrak{A} = \begin{pmatrix} s_1 \to \{m_{15}\} \\ s_2 \to \{m_{22}\} \\ s_3 \to \{m_{221}, m_{102}, m_{96}\} \\ s_4 \to \{m_{221}, m_{96}, m_{32}\} \\ s_5 \to \{m_{221}, m_{102}, m_{39}\} \\ s_6 \to \{m_{221}, m_{96}, m_{32}\} \\ s_7 \to \{m_{22}, m_{15}\} \\ s_8 \to \{m_{102}\} \\ s_9 \to \{m_{32}\} \end{pmatrix}$$

b.
$$\begin{pmatrix} m_{22} & 0.2445 \\ m_{221} & 0.1045 \\ m_{102} & 0.836 \\ m_{96} & 0.7843 \\ m_{15} & 0.1387 \\ m_{39} & 0.2383 \\ m_{32} & 0.1355 \end{pmatrix}$$

Fig. 4. (a) Meta-language of an agent in the simulation. (b) Randomly initiated reliabilities for meanings.

an agent might attach a new meaning to a particular symbol or drop a meaning from it. This requires the existence of a reliability measure for each meaning that inflects or deflects according to rewards in social interactions. Rewards are defined through explicit rules that explain the realizations of interactions. For instance, in a Prisoners Dilemma setup rewards are displayed in a tabular form. In this simulation model rewards correspond to successful interactions in terms of sense-making. If an active agent successfully makes sense to the passive agent, reliability of the mediating meaning set increases through an iterative function $r_{t+1} = r_t + e^\alpha$, where α is a positive real constant. This functional form is borrowed from the value function in Prospect Theory [31]. On the other hand our choice of this functional form is contingent to this model only and depending on the context it can be replaced with other forms.

Thus after a successful interaction, active agent inflates the reliability of the signaled meaning set and passive agent adds the meaning set into his meta-language by assigning it to an appropriate random symbol with a reliability of 0.5. Agents and meaning are recorded as an interaction. On the other hand if the result is negative, active agent deflates the reliability of the meaning set according to the formula $r_{t+1} = r_t + \lambda e^\beta$. In this model, pruning meanings is not allowed since reliability never falls to negative. However, if an agent has no meanings with reliability higher than 0.8, he deletes a random package from his meta-language and generates new ones using the algorithm in Theorem 1. Meanwhile, the passive agent does not do anything if the interaction is unsuccessful .

As the simulation runs, the number of interactions is monitored and it is observed that after 300 simulation cycles it tends to fluctuate between 500 and 700. We have reinitiated the simulation and have obtained 250 runs. In each run we have counted the number of triads formed and computed the average cultural coherence.

According to Coleman [11], fundamental indicator of embeddedness in a social group is the number of triads that is three individuals interacting with one another. Cultural coherence of two agents is defined as the ratio of the

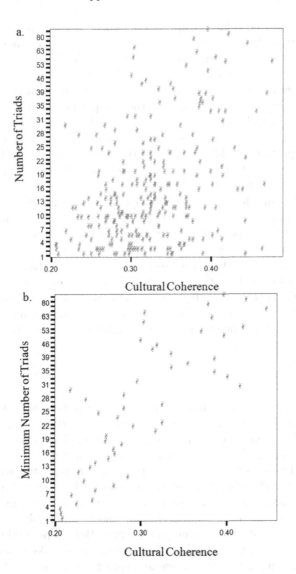

Fig. 5. (a) Number of triads and cultural coherence correspondence. (b) Left frontier representing the minimum cultural coherence sufficient to generate corresponding number of triads.

overlapping meaning sets to the count of all meanings. For instance, if the meaning collection of Agent A contains 50 distinct meanings, and meaning collection of Agent B contains 53 distinct meanings of which twenty three are common with Agent A, then the cultural coherence of these agents are computed as $0.2875 = 23/(50+53-23)$. We have computed the cultural coherence of agents in all triads and took the average.

Fig. 5a displays the relationship between cultural coherence and number of triads in 250 runs. In order to focus to minimum cultural coherence required to generate corresponding triads Fig. 5b shows only the left frontier. Results demonstrate a linear association between number of triads and minimum cultural coherence. Hence as agents interact, their meta-languages converge to each other to the extent dictated by the form of their opposition maps. In this model, agents are not allowed to prune their meaning sets albeit this could be easily implemented- since we aimed to demonstrate how large meta-language sets can get and what portion would be used effectively (reliability > 80%). Across the runs, cardinality of meta-languages ranged between 40 and 60 and the number of effective meanings were around five.

Although this simple simulation model is presented for demonstration purposes only; it implements the basic tenets of the framework we have defined so far. First, model shows a dynamic setup where agents can generate and elaborate multiplicity of meanings. Second, it exemplifies how individual mental models can interact with each other and evolve. Third, we see that a thickly coherent cultural background is not necessary for the emergence of embedded social networks, a thin coherence such as opposition maps would be sufficient to observe their dynamic formation. (Fig. 5b). Fourth, exchange of meanings through successful sense-making practices generates a social anchoring process.

4   Discussion

In this paper we have set up a framework for modeling meta-language and its implementation in a culture based simulation. We have particularly emphasized the reflexivity of the agents in terms of changing their cognitive structures according to the feedbacks from their interactions. Semiotic coherence of the agents is dynamically constructed as the interactions proceed. As a final step, we have demonstrated that embeddedness of emerging social networks is related to the semiotic coherence. Besides being a theoretical exercise for modeling sense making in a simulation model, our framework may have some interesting application potentials.

Using interview data to elicit empirical opposition maps would make a simulation model more interesting. In this regard sense-making simulations are helpful in capturing sensitivity of agents against certain inputs like policy applications. Furthermore they can inform collaboration propensities of distinct groups since common denominator of a set of individuals is not what they currently/instantaneously share but what they may share in the future.

Operationalizing culture as symbols and meanings lets us to define distinct constituents of a simulation model, such as agents, their intentions, goals and even the emerging structures to communicate with each other over the same currency; comprising a semiotic sphere. On the other hand, simulation models that adopt the cultural framework developed in this paper are not restricted to the semiotic sphere. As the rewards might realize in other spheres, the rules of the game can be represented in meta-languages. For example, consider the Prisoners' Dilemma setup. The rules of the game like selecting either to "defect" or to "cooperate" can be embedded inside meanings in meta-languages. In this way intentions will be loaded with strategies. This simple idea can be extended to games where the rules are more complicated.

References

1. Albayrak, R.S., Suerdem, A.: A Method Proposal for Elicitation of Intensional Ontological Classes: Opposition Map Approach. In: 7th International Conference on Social Science Methodology, Naples, Italy (2008)
2. Alchian, A.: Uncertainty, Evolution, and Economic Theory. The Journal of Political Economy 58, 211–221 (1950)
3. Anderson, M.L.: Embodied Cognition: A Field Guide. Artificial Intelligence 149, 91–130 (2003)
4. Berger, P.L., Luckmann, T.: The Social Construction of Reality: A Treatise in the Sociology of Knowledge. Anchor Books, New York (1966)
5. Bordini, R.H., Campbell, J.A., Vieira, R.: Extending Ascribed Intensional Ontologies with Taxonomical Relations in Anthropological Descriptions of Multi-Agent Systems. JASS 1, 4, 1–38 (1998)
6. Bourdieu, P.: The Forms of Social Capital. In: Richardson, J.G. (ed.) Handbook of Theory and Research for the Sociology of Education, pp. 241–258. Greenwood Press, New York (1986)
7. Brooks, R.: Intelligence without Representation Artificial Intelligence 47, 139–159 (1991)
8. Carley, K.: Extracting Team Mental Models through Textual Analysis. Journal of Organizational Behavior 18, 533–558 (1997)
9. Castelfranchi, C.: Through the Minds of the Agents. JASS, 1 (1998)
10. Chandler, D.: Semiotics. Routledge, New York (2002)
11. Coleman, J.S.: Social Capital in the Creation of Human Capital. American Journal of Sociology 94, 95–120 (1988)
12. Conte, R., Edmonds, B., Moss, S., Sawyer, R.K.: Computational & Mathematical Organization Theory. In: Sociology and Social Theory in AgentBased Social Simulation: A Symposium, pp. 183–205 (2001)
13. Croft, W., Cruse, D.A.: Cognitive Linguistics. Cambridge Textbooks in Linguistics. Linguistics. Cambridge University Press, Cambridge (2004)
14. Dignum, F., Dignum, V., Jonker, C.M.: Towards Agents for Policy Making. In: David, N., Sichmann, J.S. (eds.) MABS 2008. LNCS (LNAI), vol. 5269, pp. 141–153. Springer, Heidelberg (2008)
15. DiMaggio, P.: Culture and Cognition. Annual Review of Sociology 23, 263–287 (1997)
16. Dreyfus, H.: Intelligence without Representation — Merleau-Ponty's Critique of Mental Representation: The Relevance of Phenomenology to Scientific Explanation. Phenomenology and the Cognitive Sciences 1, 367–383 (2002)
17. Durkheim, E.: Durkheim, the Division of Labor in Society. The Free Press, New York (1893) (1997)
18. Edmonds, B.: Pragmatic Holism (or Pragmatic Reductionism). Foundations of Science 4, 57–82 (1999)
19. ElGuindi, F., Read, D.W.: Mathematics in Structural Theory. Current Anthropology 20, 761–773 (1979)
20. Geertz, C.: The Interpretation of Cultures. Fontana Press, London (1973)

21. Gilbert, D.T.: How Mental Systems Believe. American Psychologist 46, 107–119 (1991)
22. Gilbert, N.: Varieties of Emergence. In: Social Agents: Ecology, Exchange, and Evolution Conference Chicago (2002)
23. Gilbert, N., Troitzsch, K.G.: Simulation for the Social Scientist. Open University Press, New York (2005)
24. Goldspink, C., Kay, R.: Bridging the Micro-Macro Divide: A New Basis for Social Science. Human Relations 57, 597–618 (2004)
25. Habermas, J.: The Theory of Communicative Action. Polity, Cambridge (1984)
26. Harnad, S.: The Symbol Grounding Problem. Physica D, 335–346 (1990)
27. Hayek, F.: Individualism and Economic Order. University of Chicago Press, Chicago (1948) (1980)
28. Hofstede, G.: Cultures and Organizations: Software of the Mind. McGrow Hill, London (1991)
29. Hofstede, G.J., Jonker, C.M., Verwaart, T.: Modeling Power Distance in Trade. In: David, N., Sichmann, J.S. (eds.) MABS 2008. LNCS (LNAI), vol. 5269, pp. 1–16. Springer, Heidelberg (2008)
30. Jakobson, R., Halle, M.: Fundamentals of Language. The Hague, Mouton (1956)
31. Kahneman, D., Tversky, A.: Prospect Theory: An Analysis of Decision under Risk. Econometrica 47, 263–292 (1979)
32. Kukla, R. (ed.): Aesthetics and Cognition in Kant's Critical Philosophy. Cambridge University Press, Cambridge (2006)
33. Lakoff, G., Johnson, M.: Philosophy in the Flesh: The Embodied Mind and Its Challenge to Western Thought. Basic Books, New York (1999)
34. Loren, L.A., Dietrich, E.: Merleau-Ponty, Embodied Cognition and the Problem of Intentionality. Cybernetics and Systems 28, 345–358 (1997)
35. Lucy, J.A.: Language Diversity and Thought: A Reformulation of the Linguistic Relativity Hypothesis. Cambridge University Press, Cambridge (1996)
36. March, J.G.: A Primer on Decision Making: How Decisions Happen. The Free Press, New York (1994)
37. Merleau-Ponty, M.: Phenomenology of Perception. Routledge, London (1996)
38. Parsons, T.: An Outline of the Social System. In: Calhoun, C., Gerteis, J., Moody, J., Pfaff, S., Schmidt, K., Virk, I. (eds.) Classical Sociological Theory, pp. 366–385. Blackwell Publishers, London (2002)
39. Saussure, F.D.: Course in General Linguistics. Duckworth, London (1916) (1983)
40. Sawyer, K.: The Mechanisms of Emergence. Philosophy of the Social Sciences 34, 260–282 (2004)
41. Sawyer, R.K.: Artificial Societies: Multiagent Systems and the Micro-Macro Link in Sociological Theory. Sociological Methods Research 31 (2003)
42. Schutz, A., Luckmann, T.: The Structures of the Life-World. Heinemann, London (1973)
43. Searle, J.R.: Minds, Brains, and Programs. Behavioral and Brain Sciences 3, 417–457 (1980)
44. Sewell, W.H.: The Concept(s) of Culture. In: Bonnell, V.E., Hunt, L. (eds.) Beyond the Cultural Turn: New Directions in the Study of Society and Culture, pp. 35–62. University of California Press, Los Angeles (1999)
45. Swidler, A.: Culture in Action: Symbols and Strategies. American Sociological Review 51, 273–286 (1986)
46. Tsvetovat, M., Latek, M.: Dynamics of Agent Organizations. In: David, N., Sichmann, J.S. (eds.) MABS 2008. LNCS (LNAI), vol. 5269, pp. 60–70. Springer, Heidelberg (2008)

Author Index